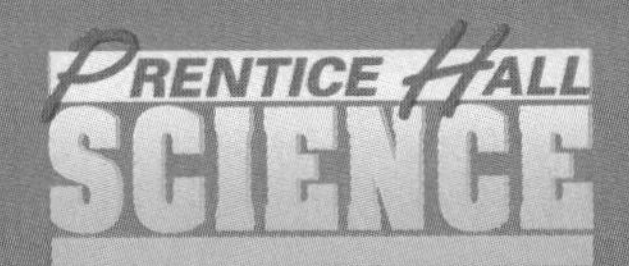

EXPLORING EARTH'S WEATHER

Anthea Maton
Former NSTA National Coordinator
Project Scope, Sequence, Coordination
Washington, DC

Jean Hopkins
Science Instructor and Department Chairperson
John H. Wood Middle School
San Antonio, Texas

Susan Johnson
Professor of Biology
Ball State University
Muncie, Indiana

David LaHart
Senior Instructor
Florida Solar Energy Center
Cape Canaveral, Florida

Charles William McLaughlin
Science Instructor and Department Chairperson
Central High School
St. Joseph, Missouri

Maryanna Quon Warner
Science Instructor
Del Dios Middle School
Escondido, California

Jill D. Wright
Professor of Science Education
Director of International Field Programs
University of Pittsburgh
Pittsburgh, Pennsylvania

Prentice Hall
Englewood Cliffs, New Jersey
Needham, Massachusetts

Prentice Hall Science

Exploring Earth's Weather

Student Text and Annotated Teacher's Edition
Laboratory Manual
Teacher's Resource Package
Teacher's Desk Reference
Computer Test Bank
Teaching Transparencies
Product Testing Activities
Computer Courseware
Video and Interactive Video

The illustration on the cover, rendered by Keith Kasnot, depicts one of Earth's most spectacular weather phenomena: a lightning storm.

Credits begin on page 136.

SECOND EDITION

ISBN 0-13-400730-1

1 2 3 4 5 6 7 8 9 10   97 96 95 94 93

Prentice Hall
A Division of Simon & Schuster
Englewood Cliffs, New Jersey 07632

STAFF CREDITS

Editorial:	Harry Bakalian, Pamela E. Hirschfeld, Maureen Grassi, Robert P. Letendre, Elisa Mui Eiger, Lorraine Smith-Phelan, Christine A. Caputo
Design:	AnnMarie Roselli, Carmela Pereira, Susan Walrath, Leslie Osher, Art Soares
Production:	Suse F. Bell, Joan McCulley, Elizabeth Torjussen, Christina Burghard
Photo Research:	Libby Forsyth, Emily Rose, Martha Conway
Publishing Technology:	Andrew Grey Bommarito, Deborah Jones, Monduane Harris, Michael Colucci, Gregory Myers, Cleasta Wilburn
Marketing:	Andrew Socha, Victoria Willows
Pre-Press Production:	Laura Sanderson, Kathryn Dix, Denise Herckenrath
Manufacturing:	Rhett Conklin, Gertrude Szyferblatt

Consultants

Kathy French	National Science Consultant
Jeannie Dennard	National Science Consultant

Contributing Writers

Linda Densman
Science Instructor
Hurst, TX

Linda Grant
Former Science Instructor
Weatherford, TX

Heather Hirschfeld
Science Writer
Durham, NC

Marcia Mungenast
Science Writer
Upper Montclair, NJ

Michael Ross
Science Writer
New York City, NY

Content Reviewers

Dan Anthony
Science Mentor
Rialto, CA

John Barrow
Science Instructor
Pomona, CA

Leslie Bettencourt
Science Instructor
Harrisville, RI

Carol Bishop
Science Instructor
Palm Desert, CA

Dan Bohan
Science Instructor
Palm Desert, CA

Steve M. Carlson
Science Instructor
Milwaukie, OR

Larry Flammer
Science Instructor
San Jose, CA

Steve Ferguson
Science Instructor
Lee's Summit, MO

Robin Lee Harris Freedman
Science Instructor
Fort Bragg, CA

Edith H. Gladden
Former Science Instructor
Philadelphia, PA

Vernita Marie Graves
Science Instructor
Tenafly, NJ

Jack Grube
Science Instructor
San Jose, CA

Emiel Hamberlin
Science Instructor
Chicago, IL

Dwight Kertzman
Science Instructor
Tulsa, OK

Judy Kirschbaum
Science/Computer Instructor
Tenafly, NJ

Kenneth L. Krause
Science Instructor
Milwaukie, OR

Ernest W. Kuehl, Jr.
Science Instructor
Bayside, NY

Mary Grace Lopez
Science Instructor
Corpus Christi, TX

Warren Maggard
Science Instructor
PeWee Valley, KY

Della M. McCaughan
Science Instructor
Biloxi, MS

Stanley J. Mulak
Former Science Instructor
Jensen Beach, FL

Richard Myers
Science Instructor
Portland, OR

Carol Nathanson
Science Mentor
Riverside, CA

Sylvia Neivert
Former Science Instructor
San Diego, CA

Jarvis VNC Pahl
Science Instructor
Rialto, CA

Arlene Sackman
Science Instructor
Tulare, CA

Christine Schumacher
Science Instructor
Pikesville, MD

Suzanne Steinke
Science Instructor
Towson, MD

Len Svinth
Science Instructor/
Chairperson
Petaluma, CA

Elaine M. Tadros
Science Instructor
Palm Desert, CA

Joyce K. Walsh
Science Instructor
Midlothian, VA

Steve Weinberg
Science Instructor
West Hartford, CT

Charlene West, PhD
Director of Curriculum
Rialto, CA

John Westwater
Science Instructor
Medford, MA

Glenna Wilkoff
Science Instructor
Chesterfield, OH

Edee Norman Wiziecki
Science Instructor
Urbana, IL

Teacher Advisory Panel

Beverly Brown
Science Instructor
Livonia, MI

James Burg
Science Instructor
Cincinnati, OH

Karen M. Cannon
Science Instructor
San Diego, CA

John Eby
Science Instructor
Richmond, CA

Elsie M. Jones
Science Instructor
Marietta, GA

Michael Pierre McKereghan
Science Instructor
Denver, CO

Donald C. Pace, Sr.
Science Instructor
Reisterstown, MD

Carlos Francisco Sainz
Science Instructor
National City, CA

William Reed
Science Instructor
Indianapolis, IN

Multicultural Consultant

Steven J. Rakow
Associate Professor
University of Houston—Clear Lake
Houston, TX

English as a Second Language (ESL) Consultants

Jaime Morales
Bilingual Coordinator
Huntington Park, CA

Pat Hollis Smith
Former ESL Instructor
Beaumont, TX

Reading Consultant

Larry Swinburne
Director
Swinburne Readability Laboratory

CONTENTS

EXPLORING EARTH'S WEATHER

Activity Bank/Reference Section

Features

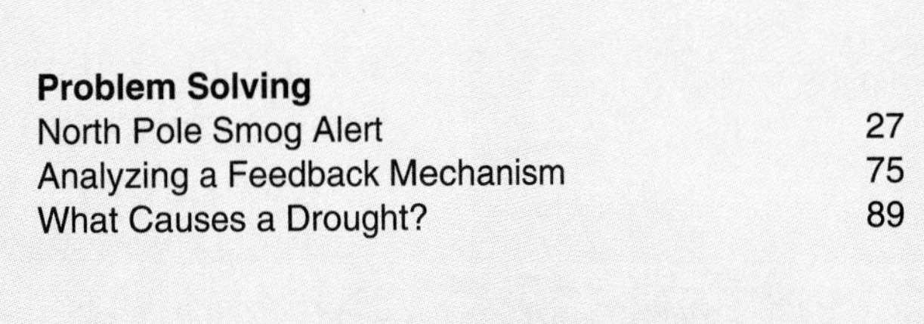

CONCEPT MAPPING

Throughout your study of science, you will learn a variety of terms, facts, figures, and concepts. Each new topic you encounter will provide its own collection of words and ideas—which, at times, you may think seem endless. But each of the ideas within a particular topic is related in some way to the others. No concept in science is isolated. Thus it will help you to understand the topic if you see the whole picture; that is, the interconnectedness of all the individual terms and ideas. This is a much more effective and satisfying way of learning than memorizing separate facts.

Actually, this should be a rather familiar process for you. Although you may not think about it in this way, you analyze many of the elements in your daily life by looking for relationships or connections. For example, when you look at a collection of flowers, you may divide them into groups: roses, carnations, and daisies. You may then associate colors with these flowers: red, pink, and white. The general topic is flowers. The subtopic is types of flowers. And the colors are specific terms that describe flowers. A topic makes more sense and is more easily understood if you understand how it is broken down into individual ideas and how these ideas are related to one another and to the entire topic.

It is often helpful to organize information visually so that you can see how it all fits together. One technique for describing related ideas is called a **concept map**. In a concept map, an idea is represented by a word or phrase enclosed in a box. There are several ideas in any concept map. A connection between two ideas is made with a line. A word or two that describes the connection is written on or near the line. The general topic is located at the top of the map. That topic is then broken down into subtopics, or more specific ideas, by branching lines. The most specific topics are located at the bottom of the map.

To construct a concept map, first identify the important ideas or key terms in the chapter or section. Do not try to include too much information. Use your judgment as to what is really important. Write the general topic at the top of your map. Let's use an example to help illustrate this process. Suppose you decide that the key terms in a section you are reading are School, Living Things, Language Arts, Subtraction, Grammar, Mathematics, Experiments, Papers, Science, Addition, Novels. The general topic is School. Write and enclose this word in a box at the top of your map.

SCHOOL

Now choose the subtopics—Language Arts, Science, Mathematics. Figure out how they are related to the topic. Add these words to your map. Continue this procedure until you have included all the important ideas and terms. Then use lines to make the appropriate connections between ideas and terms. Don't forget to write a word or two on or near the connecting line to describe the nature of the connection.

Do not be concerned if you have to redraw your map (perhaps several times!) before you show all the important connections clearly. If, for example, you write papers for Science as well as for Language Arts, you may want to place these two subjects next to each other so that the lines do not overlap.

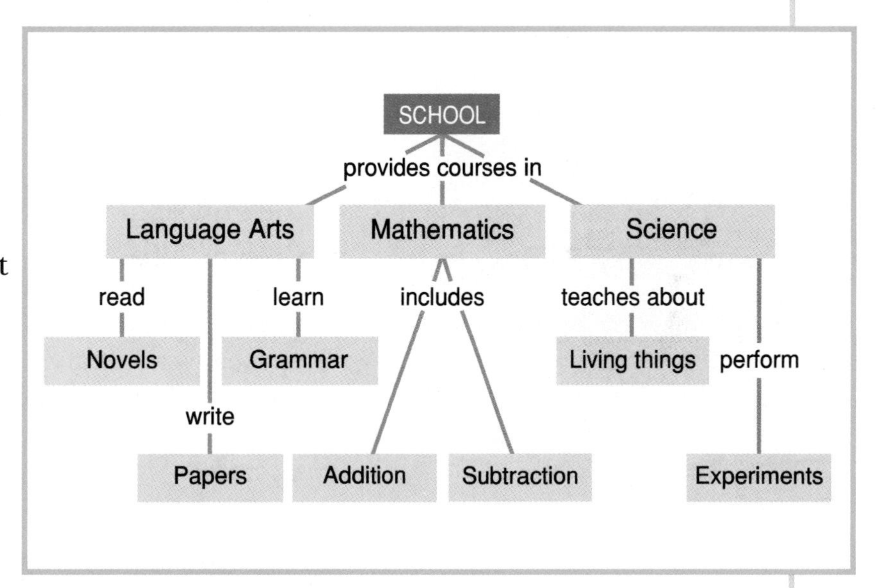

One more thing you should know about concept mapping: Concepts can be correctly mapped in many different ways. In fact, it is unlikely that any two people will draw identical concept maps for a complex topic. Thus there is no one correct concept map for any topic! Even though your concept map may not match those of your classmates, it will be correct as long as it shows the most important concepts and the clear relationships among them. Your concept map will also be correct if it has meaning to you and if it helps you understand the material you are reading. A concept map should be so clear that if some of the terms are erased, the missing terms could easily be filled in by following the logic of the concept map.

EXPLORING EARTH'S WEATHER

Weather forecasters make use of computers to coordinate data from weather satellites and ground-based observers.

Seen from space, the Earth is a "big blue marble" floating in the infinite blackness of the universe. The first astronauts to travel into space looked back at the Earth and were struck by the delicate beauty of their home planet—a tiny oasis of life in a hostile universe. As one astronaut said: "I was terrified by its fragile appearance."

For the first time, the astronauts saw the Earth's atmosphere—the "ocean of air"—as a thin blue ribbon surrounding the planet. They found that bands of white storm clouds swirling above dark blue oceans were the most visible features on the Earth's surface. The state of the atmosphere, as shown by the presence of these clouds, represents the Earth's weather. In this book you will learn about weather and how it affects your life. You will find out what causes weather and what people can do to predict (and perhaps control) it. You will also learn about long-term weather conditions, or climate.

The destructive power of a hurricane can clearly be seen in the damage caused to Galveston, Texas, by Hurricane Alicia.

CHAPTERS

Although climate seems to remain about the same from year to year, it is actually changing slowly over a period of thousands or even millions of years. What causes changes in climate? Are humans speeding up these climate changes? You will learn the answers to these and other questions you may have as you read the chapters that follow. And you will also learn how climate influences life—plants, animals, and people—on Planet Earth.

A winter snowstorm and a scorching day at the beach illustrate contrasting climates in different parts of the United States.

Discovery Activity

Weather Watch

Look out your classroom window. Is the sun shining? Are there clouds in the sky? Is it raining or snowing? Every day that you use this textbook, describe the weather where you live. Record the temperature, precipitation, and any other weather factors that are important to you.

- Based on your observations, what can you predict about the weather the week after you finish this textbook? How about the following year?

What Is Weather?

Guide for Reading

After you read the following sections, you will be able to

1–1 Heating the Earth
- Identify the factors that interact to cause weather.

1–2 Air Pressure
- Describe the three factors that affect air pressure.

1–3 Winds
- Explain the difference between local and global wind patterns.

1–4 Moisture in the Air
- Identify the three basic types of clouds.

1–5 Weather Patterns
- Describe how fronts affect weather patterns.

1–6 Predicting the Weather
- Explain how weather forecasts are made.

The animal in the photograph is a groundhog. You might be wondering what a groundhog has to do with the weather. Well, many people believe that a groundhog can forecast the weather! In fact, in most countries of the Northern Hemisphere, February 2 is called Groundhog Day. Early on Groundhog Day, people anxiously wait for a groundhog to come out of its burrow. They believe that if the groundhog sees its shadow, six more weeks of winter are to come.

The belief that groundhogs can forecast the weather is just one example of weather folklore. You may have heard the weather proverb: "Red sky at night, sailors delight. Red sky at morning, sailors take warning." Perhaps you can think of other examples of weather folklore. Such folklore is common because weather influences every aspect of our lives.

People have been trying to make accurate weather predictions for centuries. Today, scientists know a great deal about the conditions that influence weather. Modern instruments help them predict the weather more accurately than ever before. In this chapter you will learn about weather and how scientists use weather satellites and computers—not groundhogs—to make their forecasts.

Journal *Activity*

You and Your World Have you ever experienced a violent storm, such as a hurricane or tornado? What was the storm like? How did you feel during the storm? Did the storm cause any damage? What questions came to mind as you observed the storm? In your journal, describe your favorite storm story.

If a groundhog sees its shadow when it emerges from its burrow, weather folklore says there will be six more weeks of winter.

Guide for Reading

Focus on these questions as you read.

- *What factors interact to cause weather?*
- *How is the Earth's atmosphere heated?*

1–1 Heating the Earth

When you woke up this morning, did you stop to think about the weather? Was the sun shining? Was it warm enough for a picnic? Did you take your umbrella with you?

Weather affects your daily life and influences you and the world around you. The type of homes people build, the clothes they wear, the crops they grow, the jobs they perform, and the ways in which they spend their leisure time are all determined by the weather.

Today, people have a good understanding of the weather. Weather satellites, computers, and other kinds of weather instruments provide accurate information about weather conditions. Meteorologists (meet-ee-uh-RAHL-uh-jihsts), people who study the weather, use this information to predict the weather. Their forecasts help you plan your daily activities. But what exactly is weather and what causes it?

You can think of weather as the daily condition of the Earth's **atmosphere** (AT-muhs-feer). The atmosphere is a mixture of gases that surround the Earth. Weather is caused by the interaction of several factors in the atmosphere. **The atmospheric factors that interact to cause weather are heat energy, air pressure, winds, and moisture.**

Figure 1–1 *Weather plays an important role in our daily lives—from the clothes we wear to the crops we grow.*

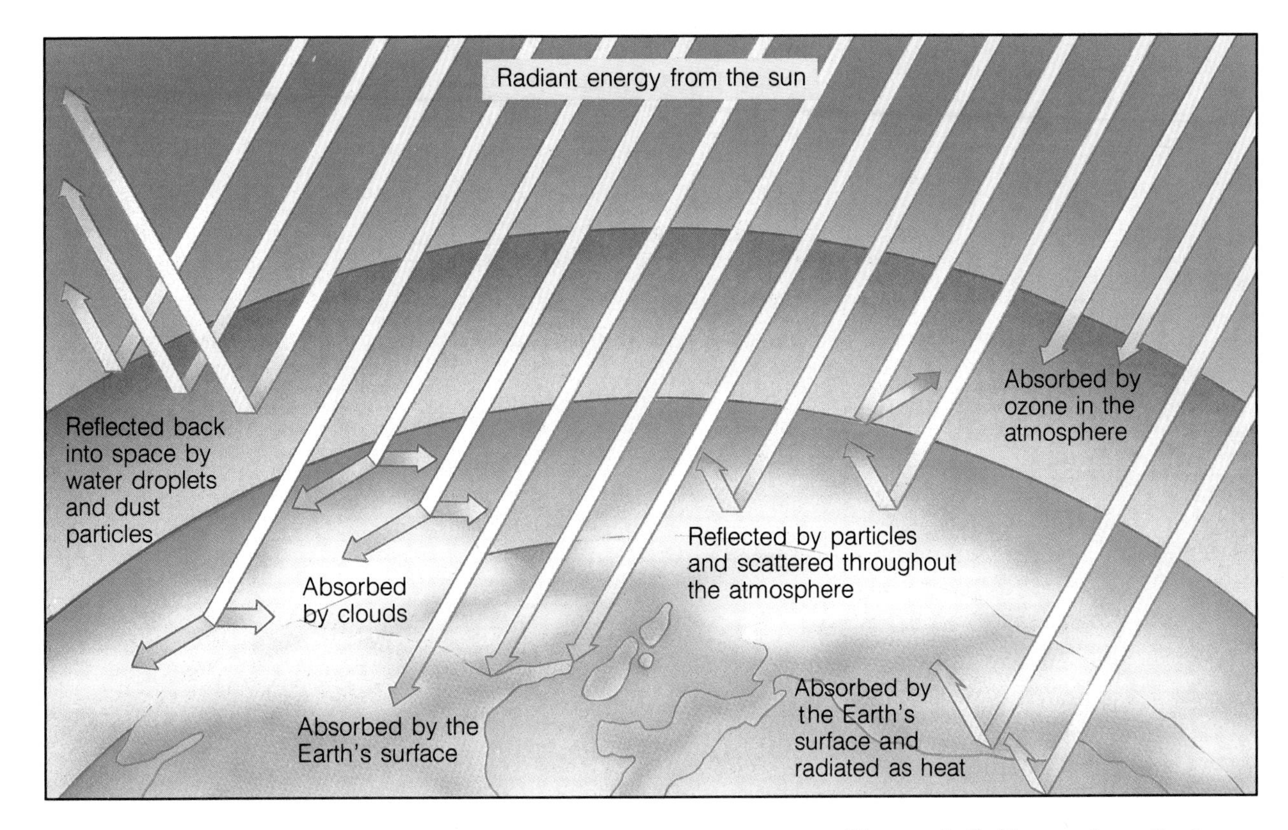

Figure 1–2 *The sun's radiant energy helps warm the atmosphere. The atmosphere then warms the Earth. According to the diagram, what happens to the sun's radiant energy?*

Heat Energy and the Atmosphere

Almost all of the Earth's energy comes from the sun. This energy is called radiant energy. The sun's radiant energy warms the Earth. The atmosphere also helps warm the Earth by absorbing, storing, and recycling the sun's radiant energy. Let's see how this happens.

As the sun's energy reaches the atmosphere, part of it is reflected (bounced back) into space and part is scattered throughout the atmosphere. This happens when incoming rays of sunlight strike water droplets and dust particles in the atmosphere.

Much of the sun's energy that is scattered throughout the atmosphere is absorbed by the atmosphere. In the upper atmosphere, a layer of ozone gas (O_3) absorbs one form of radiant energy called ultraviolet radiation. You have probably heard that ultraviolet radiation, which causes sunburn, can be dangerous to people. Too much ultraviolet radiation can cause skin cancer. That is why the ozone layer, which absorbs much of the ultraviolet radiation from the sun, is so important to life on Earth. Ultraviolet radiation does, however, have some

How Can You Prevent Food From Spoiling?, p.120

Convection Currents

1. Pour water into a small beaker until it is almost full.

2. Add two or three drops of food coloring to the surface of the water.

3. Put on safety goggles and light a candle. **CAUTION:** *Be careful when working near an open flame.*

4. Put on heat-resistant gloves. Using laboratory tongs, hold the beaker about 10 centimeters above the candle flame.

What happens to the food coloring? Why? How are your observations related to the heating of the air? At what location on the Earth would air constantly be rising because of convection currents?

beneficial uses. Ultraviolet lamps are used to kill bacteria in hospitals and in food-processing plants, where bacteria could cause packaged foods to spoil.

Radiant energy that is neither reflected nor absorbed by the atmosphere reaches the Earth's surface. Here it is absorbed by the Earth and changed into heat. **The sun's energy that is absorbed by the Earth is spread throughout the atmosphere in three basic ways: conduction, convection, and radiation.**

Heat Transfer in the Atmosphere

Conduction is the direct transfer of heat energy from one substance to another. As air above the Earth's surface comes into contact with the warm ground, the air is warmed. So temperatures close to the ground are usually higher than temperatures a few meters above the ground. However, soil, water, and air are poor conductors of heat. So conduction plays only a minor role in heating the land, ocean, and atmosphere.

Convection is the transfer of heat energy in a fluid (gas or liquid). Air is a fluid. When air near the Earth's surface is heated, it becomes less dense and rises. Cooler, denser air from above sinks. As the cool air sinks, it is heated by the ground and begins to rise. This process of warm air rising and cool air sinking forms convection currents. Convection currents are caused by the unequal heating of the

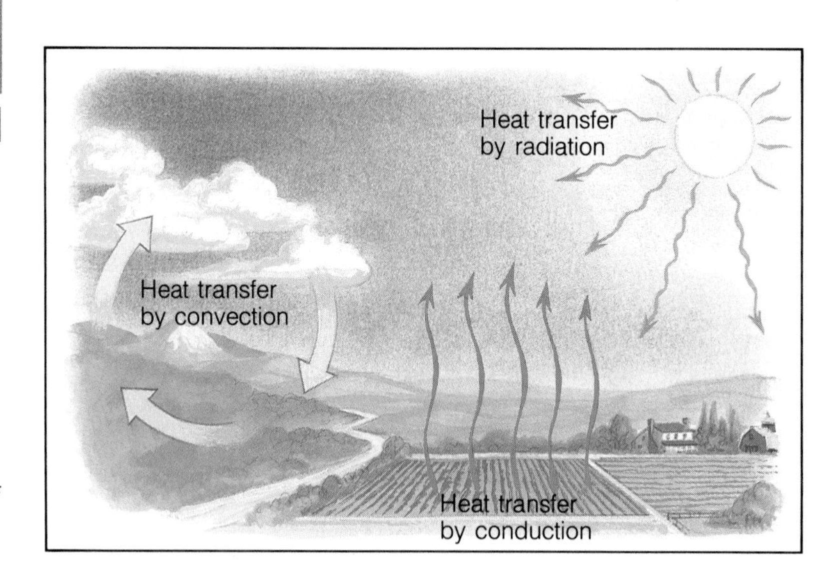

Figure 1–3 *Conduction is the direct transfer of heat energy from one substance to another. Convection is the transfer of heat in a fluid. Radiation is the transfer of heat energy through space in the form of waves. How is most of the heat energy in the atmosphere transferred?*

atmosphere. Most of the heat energy in the atmosphere is transferred by convection currents.

Radiation is the transfer of heat energy through empty space. Heat energy that is transferred by radiation does not need the presence of a solid, liquid, or gas. It can travel through a vacuum, or empty space. Heat from the sun reaches the Earth by radiation. When radiant energy from the sun is absorbed by the Earth, it is changed into heat.

The Greenhouse Effect

As you have just read, some of the sun's radiant energy (in the form of ultraviolet rays) is absorbed by the Earth and changed into heat. Ultraviolet rays pass easily through the atmosphere and reach the Earth. Later, this energy is radiated back from the Earth to the atmosphere in the form of infrared rays. You cannot see infrared rays, but you can feel them as heat. (Although humans cannot see infrared rays, rattlesnakes and some other snakes have heat-sensitive pits on their head that "see" the heat given off by small animals.)

You may actually be more familiar with infrared rays than you realize. The heat lamps often used in

Figure 1–4 *In a greenhouse, the glass windows prevent heat from escaping (left). Carbon dioxide and other gases in the atmosphere act like the glass windows in a greenhouse (right). In what form is the energy of ultraviolet rays radiated from the Earth to the atmosphere?*

Figure 1–5 *Rattlesnakes have special heat-detecting organs. These pit organs are so sensitive they can detect differences in temperature of only 0.003°C!*

Figure 1–6 *The extremely high surface temperatures on Venus are the result of a runaway greenhouse effect. Carbon dioxide released by burning fossil fuels may already be causing higher temperatures on Earth.*

restaurants to keep food warm make use of infrared radiation. If you hold your hand near a light bulb or stove, you can feel the heat given off as infrared rays.

Infrared rays are not like ultraviolet rays, however. Infrared rays cannot pass through the atmosphere and out into space. Carbon dioxide (CO_2) and other gases in the atmosphere absorb the infrared rays, forming a kind of "heat blanket" around the Earth. This process is called the **greenhouse effect** because the carbon dioxide acts like the glass in a greenhouse to trap heat. The greenhouse effect makes the Earth a comfortable place to live. What do you think would happen to the temperature at the Earth's surface if there were no greenhouse effect?

Because most of the infrared rays are absorbed by carbon dioxide, the amount of this gas in the atmosphere is very important. Carbon dioxide is produced by burning fossil fuels, such as coal, oil, and natural gas. As the amount of carbon dioxide in the atmosphere increases, more infrared rays will be absorbed. The greenhouse effect will increase and temperatures at the Earth's surface will go up.

Higher temperatures might result in altered weather patterns, including warmer winters and changes in rainfall. You will learn more about the results of the greenhouse effect in Chapter 2.

Temperature Variations

If the Earth's atmosphere is warmed by heat rising from the surface, how can the air temperature vary so much from place to place? To help you answer this question, look at Figure 1–7.

The angle at which the sun's rays strike the surface is not the same everywhere on Earth. At the equator (the imaginary line that separates the Earth into two halves), the sun is nearly overhead. The sun's rays strike the Earth at a 90° angle all year long. The greatest heating occurs where the sun's rays are most direct; that is, at or near an angle of 90°. So areas at or near the equator receive the most radiant energy and have the highest temperatures.

The farther away from the equator an area is, the less radiant energy it receives. Why is this so? In these areas, the angle at which the sun's rays strike the Earth is less than 90°. As the angle of the sun's

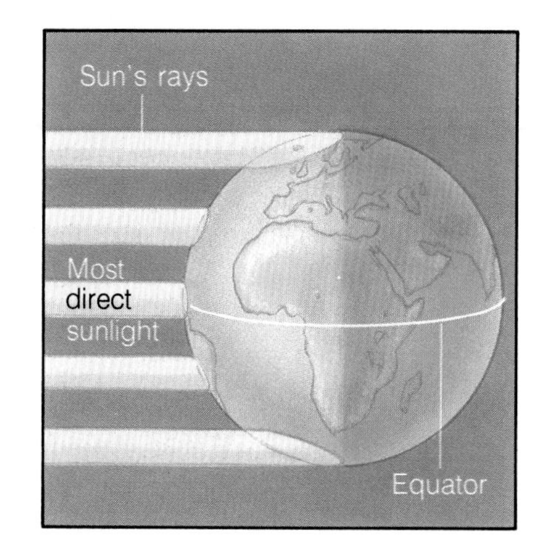

Figure 1–7 *Radiant energy from the sun strikes the Earth at different angles, causing uneven heating of the Earth's surface. Which area receives the most direct sunlight? In which areas is the same amount of radiant energy spread over a wider area?*

Figure 1–8 *What is responsible for the wide variations in air temperature in different regions of the Earth?*

rays becomes smaller, the rays become less direct. The same amount of radiant energy is spread over a wider area. The result is less heat and lower temperatures.

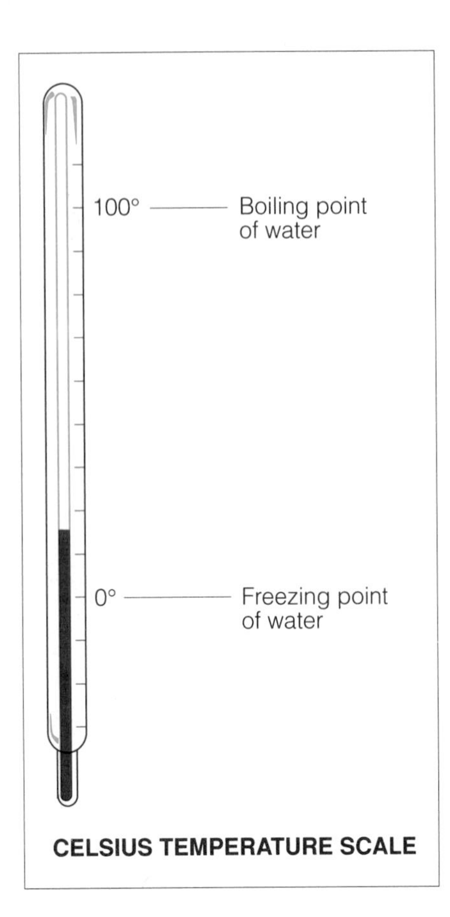

Figure 1–9 *The temperature of the air is measured with a thermometer in units called degrees. What is the freezing point of water on the Celsius scale? The boiling point?*

Measuring Temperature

Air temperature is measured with a **thermometer.** Most thermometers consist of a thin glass tube with a bulb at one end. The bulb is filled with a liquid, usually either mercury or alcohol that is colored with dye.

Thermometers make use of the ability of a liquid to expand and contract. When a liquid is heated, it expands, or takes up more space. When a liquid is cooled, it contracts, or takes up less space. What happens to the liquid in a thermometer when the air temperature rises? What happens to the liquid when the air temperature falls?

Temperature is measured in units called degrees (°). The temperature scale used by scientists is the Celsius (SEHL-see-uhs) scale. On the Celsius scale, the freezing point of water is 0°. The boiling point of water is 100°C. Normal human body temperature is 37°C.

1–1 Section Review

1. What are the factors that interact to cause weather?
2. What are three ways by which heat energy is spread throughout the atmosphere?
3. How does the angle at which the sun's rays strike the Earth affect the temperature at the Earth's surface?
4. How is carbon dioxide gas in the atmosphere similar to the glass in a greenhouse?

Connection—*Environmental Science*

5. Why do you think some scientists are concerned about increased levels of carbon dioxide in the atmosphere as a result of burning fossil fuels? What effect might an increase in carbon dioxide have on the environment?

1–2 Air Pressure

Hold the eraser of a pencil against the palm of your hand. Now press down. What you feel is the force of the pencil pressing against your hand. You feel pressure. Atmospheric pressure, or **air pressure,** is a measure of the force of the air pressing down on the Earth's surface. The air pressure at any point on the Earth is equal to the weight of the air directly above that point. We are walking on the bottom of an "ocean" of air about 800 kilometers deep!

Air pressure can vary from one point to another on the Earth's surface. **The air pressure at any particular point on the Earth depends on the density of the air.** (Density is equal to mass divided by volume.) Denser air has more mass per unit volume than less dense air. So denser air exerts more air pressure against the Earth's surface than less dense air does.

Guide for Reading

Focus on this question as you read.

- *What is the relationship between the density of air and air pressure?*

Factors That Affect Air Pressure

The density of the Earth's atmosphere, and thus air pressure, is affected by three factors: temperature, water vapor, and elevation. As you learned in Section 1–1, the density of a fluid (gas or liquid) decreases when the fluid is heated. Less dense air exerts less air pressure. So places with high

Figure 1–10 *Elevation, or altitude, is one of the factors that affect air pressure. Is the air pressure at the top of Mt. McKinley, Alaska, higher or lower than the air pressure at sea level?*

AIR PRESSURE AND ALTITUDE

Altitude (m)	Air Pressure (g/cm^2)
Sea level	1034
3000	717
6000	450
9000	302
12,000	190
15,000	112

A Water Barometer

Mercury has a density of 13.5 g/cm^3. If standard air pressure supports a column of mercury 76 cm high, how high could the column of water rise when supported at this pressure? (The density of water is 1 g/cm^3.)

temperatures usually have lower air pressure than places with low temperatures.

Warmer, less dense air can hold more water vapor than colder, denser air. So as the amount of water vapor in the air increases, the mass of the air decreases and the air becomes less dense. Thus air with a large amount of water vapor in it exerts less air pressure than dryer air.

Elevation, or altitude, also affects air pressure. As the elevation (height above sea level) increases, the air becomes thinner, or less dense. So the air pressure decreases with increasing elevation.

Measuring Air Pressure

Because air pressure changes with changes in temperature and elevation, standard air pressure is measured at a temperature of 0°C at sea level. Air pressure is measured with an instrument called a **barometer** (buh-RAHM-uh-ter). There are two types of barometers. One type, called a mercury barometer, is shown in Figure 1–11.

The mercury barometer was invented in 1643 by an Italian scientist named Evangelista Torricelli. A mercury barometer consists of a glass tube closed at one end and filled with mercury. The open end of the glass tube is placed in a container of mercury. At sea level, air pressure pushing down on the surface of the mercury in the container supports the column of mercury at a certain height in the tube. As the air pressure decreases, the column of mercury drops. What will happen to the column of mercury if the air pressure increases?

A more common type of barometer, called an aneroid (AN-uh-roid) barometer, is shown in Figure 1–12. An aneroid barometer consists of an airtight metal box from which most of the air has been removed. (The word aneroid comes from a Greek word meaning without liquid.) A change in air pressure causes a needle to move along a dial, which indicates the new air pressure.

Figure 1–11 *When air pressure increases, the column of mercury rises in the barometer tube (right). What happens to the column of mercury when air pressure decreases (left)?*

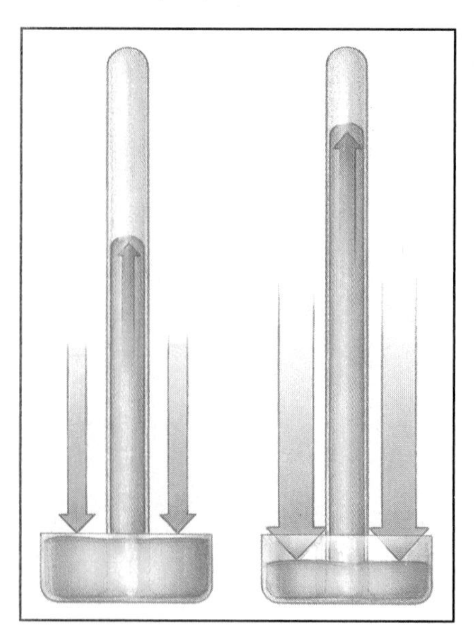

Air Pressure and Weather

Barometers can be used to help forecast the weather. Air pressure may become relatively high when large masses of air come together in the upper

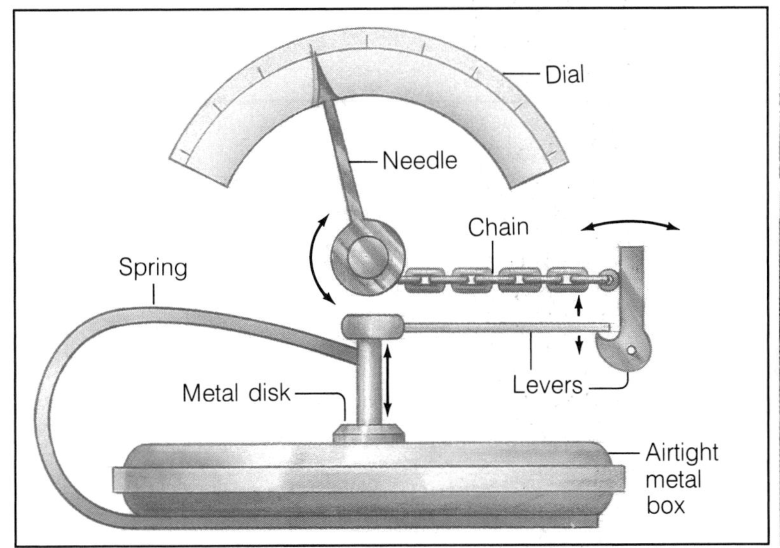

Figure 1–12 *An aneroid barometer is often used in homes, offices, and classrooms to detect changes in air pressure.*

atmosphere. These air masses press down on the layers of air below. This pressure usually prevents warm, moist air from rising into the upper atmosphere. As a result, clouds do not form. So high pressure usually means fair weather. But there are exceptions.

Air pressure may become relatively low when large air masses move apart in the upper atmosphere. This reduces pressure on the layers of warm air below. As a result, the warm air rises. If the warm air is moist, clouds will form in the upper atmosphere. So low pressure can lead to cloudy, rainy weather. But, again, there are exceptions.

1–2 Section Review

1. What is air pressure? What is the relationship between the density of air and air pressure?
2. List and describe three factors that affect air pressure.
3. How is air pressure measured?

Critical Thinking—*Relating Concepts*

4. Which of the following cities has the highest air pressure? Which has the lowest? Explain.

Atlanta, Georgia	elevation: 320 meters
Boise, Idaho	elevation: 825 meters
Denver, Colorado	elevation: 1600 meters
Salt Lake City, Utah	elevation: 1300 meters

ACTIVITY
DISCOVERING

Using a Barometer to Forecast the Weather

1. For five days, use a barometer to measure air pressure. Make your measurements before school, during science class, and after school. Record your measurements.

2. At the same time, observe and record the weather conditions.

3. Make a graph of your air pressure measurements.

■ Did you see any relationship between your air pressure measurements and the weather conditions? Can a barometer help you to predict the weather? Explain.

Guide for Reading

Focus on this question as you read.

- *What is the origin and movement of local winds and global winds?*

1–3 Winds

Have you ever flown a kite at the beach? A beach is a good place to fly a kite because of the winds that usually blow near the shore. What causes these winds to blow? When air is heated, its density decreases. The warm air rises and produces an area of low pressure. Cooler, denser air, which produces an area of high pressure, moves in underneath the rising warm air. So air moves from an area of high pressure to an area of lower pressure. **Winds** are formed by this movement of air from one place to another.

There are two general types of winds: local winds and global winds. Local winds are the type you are most familiar with. They blow from any direction and usually cover short distances. Global winds blow from a specific direction and almost always cover longer distances than local winds. **Both local winds and global winds are caused by differences in air pressure due to unequal heating of the atmosphere.**

The Density of Water

Is cold water denser than hot water? Try this activity to find out.

1. Fill a deep pan three-fourths full of cold water.

2. Fill a small bottle with hot (not boiling) water. Add a few drops of food coloring to the hot water.

3. Hold your finger over the opening of the bottle. Carefully place the bottle on its side in the pan of cold water. Make sure the bottle is completely under water.

4. Take your finger away from the opening of the bottle. Observe what happens.

What happened when you removed your finger? What does this tell you about the density of hot water and cold water?

■ What do you think would happen if you put hot water in the pan and cold water in the bottle? Try it and find out.

Local Winds

During the day, the air over a land area is often warmer than the air over a nearby lake or sea. The air is warmer because the land heats up faster than the water. As the warm air over the land rises, the cooler air over the sea moves inland to take its place.

Figure 1–13 *Wind is air in motion. The force of the wind enables the crew of this racing yacht to enjoy an exciting ride.*

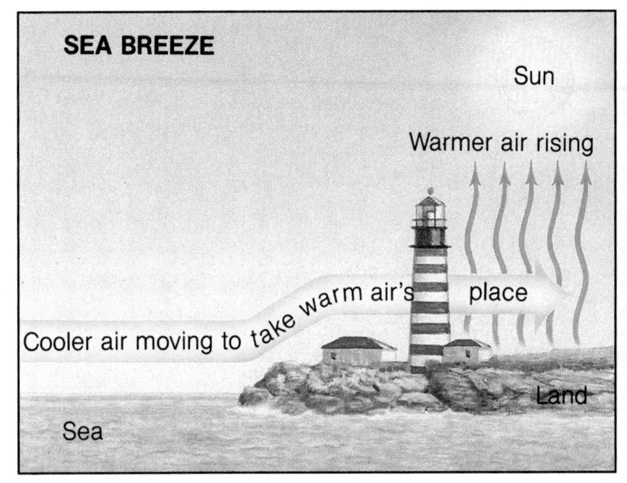

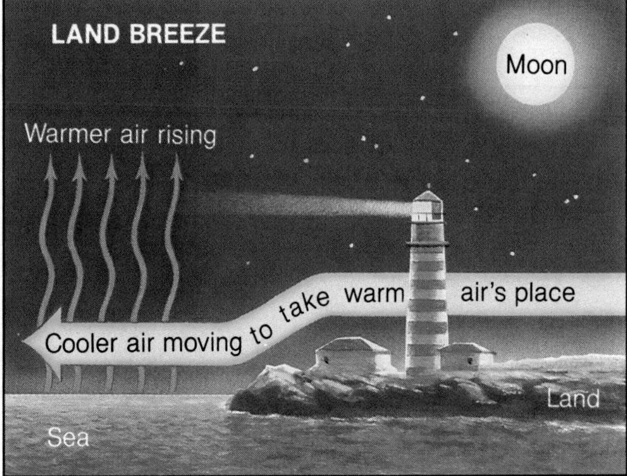

Figure 1–14 *Land and water absorb and lose heat at different rates, causing a sea breeze during the day and a land breeze during the night. Which heats up faster: land or water?*

This flow of air from the sea to the land is called a **sea breeze.** If you have ever spent a summer's day at the beach, you have probably felt a sea breeze.

During the night, the land cools off faster than the water. The air over the sea is now warmer than the air over the land. This warm air over the sea rises. The cooler air over the land moves to replace the rising warm air over the sea. A flow of air from the land to the sea, called a **land breeze,** is formed. If you have stayed at the beach after sunset, then you are probably familiar with a land breeze, too. A land breeze is also called an off-shore breeze.

The name of a wind tells you from which direction the wind is blowing. A land breeze blows from the land to the sea. A sea breeze blows from the sea to the land. Most local winds that you are familiar with are named according to the direction from which they are blowing. For example, a northwest wind blows from northwest to southeast. From what direction does a southwest wind come? In what direction is it blowing?

A major land and sea breeze is called a monsoon (mahn-SOON). A monsoon is a seasonal wind. (The word monsoon is derived from an Arabic word that means season.) During part of the year, a monsoon blows from the land to the ocean. During the rest of the year, it blows from the ocean to the land. When a monsoon blows from the ocean to the land, it brings in warm, moist air. This results in a rainy season with warm temperatures and huge amounts of rain. The rainy season is important to many countries because it supplies the water needed for farming. Monsoon winds are very common in Asia.

Heating Land and Water

Obtain the following materials: two beakers, sand, water, a thermometer, a watch or clock, and a bright light bulb (or a sunny window). Using these materials, design an experiment to answer these questions.

1. Which heats up faster: land or water?

2. Which one cools down faster?

3. Which one holds heat longer?

■ Based on the results of your experiment, explain why land and sea breezes occur.

Figure 1–15 *Because of the Earth's rotation, winds appear to curve to the right in the Northern Hemisphere and to the left in the Southern Hemisphere. What is the name for this shift in wind direction?*

Global Winds

Unequal heating of the Earth's surface also forms large global wind systems. In areas near the equator, the sun is almost directly overhead for most of the year. The direct rays of the sun heat the Earth's surface rapidly. The polar regions receive slanting rays from the sun. The slanting rays do not heat the Earth's surface as rapidly as the direct rays do. So temperatures near the poles are lower than those near the equator. At the equator, the warm air rises and moves toward the poles. At the poles, the cooler air sinks and moves toward the equator. This movement produces a global pattern of air circulation.

Global winds do not move directly from north to south or from south to north as you might expect. Because the Earth rotates, or spins on its axis, from west to east, the paths of the winds shift in relation to the Earth's surface. All winds in the Northern Hemisphere curve to the right as they move. In the Southern Hemisphere, winds curve to the left. This shift in wind direction is called the **Coriolis effect.**

The Coriolis effect is the apparent shift in the path of any fluid or object moving above the surface of the Earth due to the rotation of the Earth. For example, suppose you are in an airplane flying south from Seattle, Washington, to San Jose, California. If the pilot does not adjust for the Coriolis effect, the airplane will land west of the point for which it is headed. In other words, an invisible force seems to be pushing the airplane west. You might wind up in the Pacific Ocean!

The diagram in Figure 1–16 shows the Earth's global wind systems. Refer to it often as you read the description of each global wind system. Remember, wind systems describe an overall pattern of air movement. At any particular time or place, local conditions may influence and change the pattern.

DOLDRUMS At the equator (0° latitude), surface winds are quite calm. These winds are called the doldrums (DOHL-druhmz). A belt of air around the equator receives much of the sun's radiant energy. The warm, rising air produces a low-pressure area that extends many kilometers north and south of the equator. Cooler, high-pressure air would normally flow into such an area, creating winds. But the

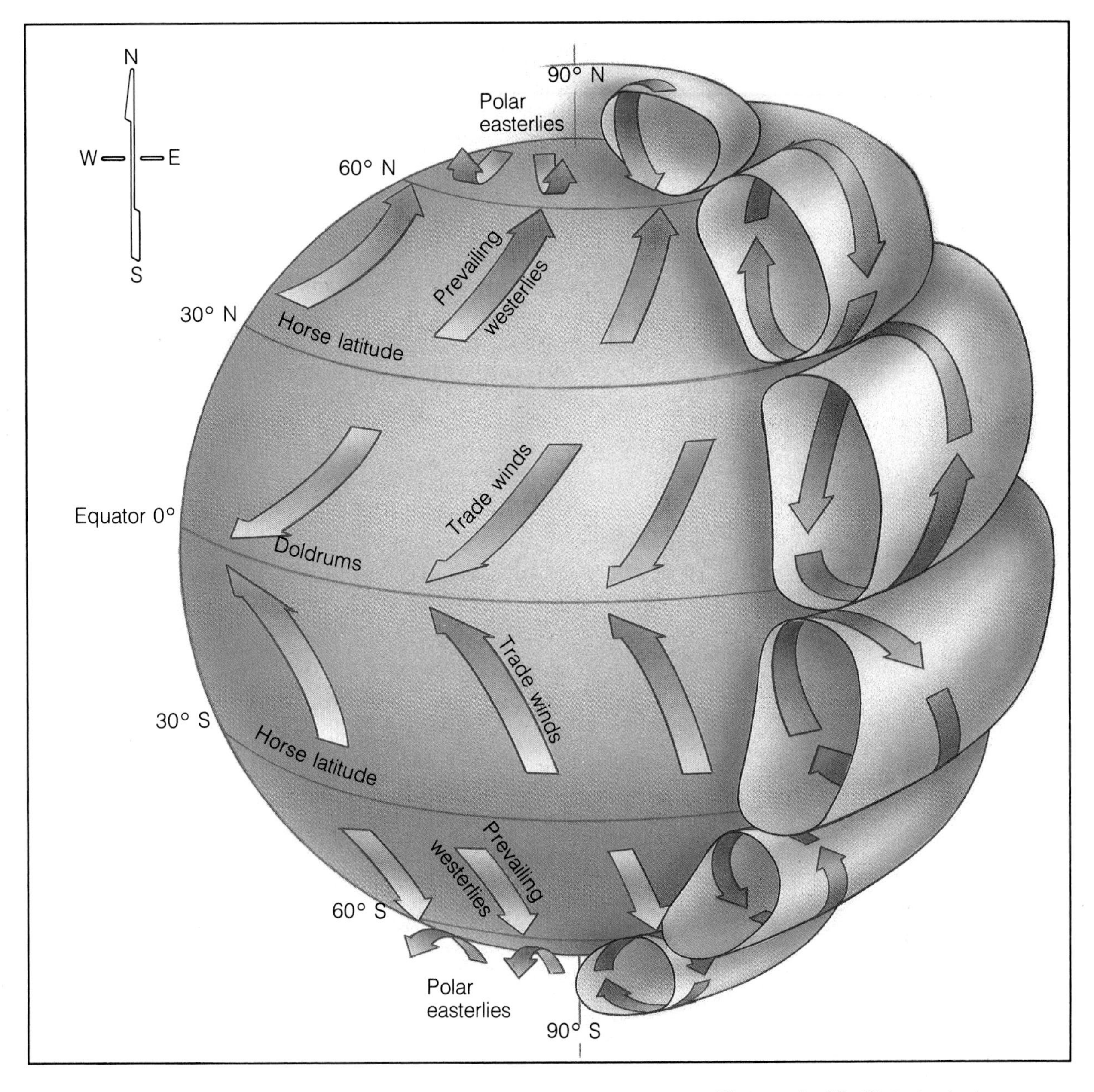

Figure 1–16 *Global wind patterns are caused by the unequal heating of the Earth's surface and by the rotation of the Earth. Warm air rises, cold air sinks, and the Coriolis effect causes the winds to curve. What are the three major global winds?*

cooler air is warmed so rapidly near the equator that the winds which form cannot move into the low-pressure area. As a result, any winds that do form are weak. The doldrums can be a problem for sailing ships. Because there may be no winds, or weak winds at best, sailing ships can be stuck in the doldrums for many days. Have you ever heard people refer to themselves as being "in the doldrums"? What did they mean?

TRADE WINDS About 30° north and south of the equator, the warm air rising from the equator cools and begins to sink. Here, the sky is usually clear.

Figure 1–17 *The overall movement of global wind systems can be seen in the pattern of the Earth's cloud cover.*

There are few clouds and little rainfall. Winds are calm. Hundreds of years ago, sailing ships traveling to the New World were sometimes unable to move for days or weeks because there was too little wind. Sailors sometimes had to throw horses overboard when the horses' food supply ran out. For this reason the latitudes 30° north and south of the equator are called the horse latitudes.

At the horse latitudes, some of the sinking air travels back toward the equator. The rest of the sinking air continues to move toward the poles. The air moving back toward the equator forms a belt of warm, steady winds. These winds are called trade winds.

In the Northern Hemisphere, the Coriolis effect deflects the trade winds to the right. These winds, called the northeast trades, blow from northeast to southwest. In the Southern Hemisphere, the trade winds are deflected to the left. They become the southeast trades. In what direction do the southeast trades blow?

Early sailors used the trade winds when they traveled to the New World. The trade winds provided a busy sailing route between Europe and America. Today, airplane pilots use the trade winds to increase speed and save fuel when they fly this route from east to west.

Figure 1–18 *Unlike the trade winds, the prevailing westerlies are strong winds. Where are the prevailing westerlies located?*

PREVAILING WESTERLIES The cool, sinking air that continues to move toward the North and South poles is also influenced by the Coriolis effect. In the Northern Hemisphere, the air is deflected to the right. In the Southern Hemisphere, it is deflected to the left. So in both hemispheres, the winds appear to travel from west to east. These winds are called the prevailing westerlies. (Remember, winds are named according to the direction from which they blow.) As you can see from Figure 1–16, the prevailing westerlies are located in a belt from 40° to 60° latitude in both hemispheres. Unlike the trade winds, the prevailing westerlies are often particularly strong winds.

POLAR EASTERLIES In both hemispheres, the westerlies start rising and cooling between 50° and 60° latitude as they approach the poles (90° latitude). Here they meet extremely cold air flowing toward the equator from the poles. This band of cold air is deflected west by the Coriolis effect. As a result, the winds appear to travel from east to west and are called the polar easterlies. The polar easterlies are cold but weak winds. In the United States, many changes in the weather are caused by the polar easterlies.

Figure 1–19 *A high-altitude jet stream moves over the Nile River Valley and the Red Sea. In which direction do jet streams flow?*

Jet Streams

For centuries, people have been aware of the global winds you have just read about. But it was not until the 1940s that another global wind was discovered. This wind is a narrow belt of strong, high-speed, high-pressure air called a jet stream. Jet streams flow from west to east at altitudes above 12 kilometers. Wind speeds in the jet streams can reach 180 kilometers per hour in the summer and 220 to 350 kilometers per hour in the winter. Airplane pilots flying from west to east can use a jet stream to increase speed and save fuel.

Problem Solving

North Pole Smog Alert

No place on the surface of the Earth is farther away from industry and human development than the land above the Arctic Circle near the North Pole. Yet scientists have discovered sulfur particles in the arctic air that are identical to those found in the polluted air of some European cities. These particles are so thick that they form a blue-gray haze similar to that seen over many large cities. Using your knowledge of global wind systems, explain how air pollution has reached the Arctic. Draw a map to illustrate your explanation.

Jet streams do not flow around the Earth in regular bands. They wander up and down as they circle the Earth. At times, they take great detours north and south. The wind speed and depth of a jet stream can change from season to season, or even from day to day.

The wandering jet streams affect the atmosphere below them. The rush of a jet stream creates waves and eddy currents, or swirling motions opposite to the flow of the main stream, in the lower atmosphere. These disturbances cause air masses in the lower atmosphere to spread out. This produces areas of low pressure. The low-pressure areas serve as the centers of local storms.

Figure 1–20 *Wind speed is related to the rate at which the cups of the anemometer revolve. What other weather instrument can you see in the photograph?*

Build Your Own Anemometer, p.121

Measuring Wind

As you have been reading about local and global winds, you have probably noticed that two measurements are needed to describe wind: wind direction and wind speed. Meteorologists and weather observers use a wind vane to determine the direction of the wind on the Earth's surface. A wind vane points into the wind. An **anemometer** (an-uh-MAHM-uh-ter) is used to measure wind speed. Wind speed is usually expressed in meters per second, miles per hour, or knots. One knot is equal to 1850 meters per hour.

ACTIVITY DOING

Build a Wind-Speed Meter

1. Obtain a square piece of cardboard.

2. Stick a push pin into the upper left corner.

3. Hang a 3-cm strip of metal from the push pin.

4. Calibrate your wind-speed meter by holding it out the window of a moving car. The metal strip will move higher as the wind speed increases. Mark the position of the strip at different speeds.

1–3 Section Review

1. What are the differences between local winds and global winds? How are they alike?
2. What causes winds in the Northern Hemisphere to curve to the right as they move?
3. Name the Earth's four major wind belts.
4. Describe the movements of the three major global winds in terms of unequal heating and the Coriolis effect.

Connection—*You and Your World*

5. An airplane trip from New York City to Los Angeles takes longer than the return trip from Los Angeles to New York. Explain why.

1–4 Moisture in the Air

Guide for Reading

Focus on these questions as you read.

- *How are the terms humidity and relative humidity related?*
- *What are the forms of water vapor in the air?*

As you walk to the supermarket on a summer afternoon, you can feel your shirt sticking to your back. Beads of salty perspiration cling to your forehead and upper lip. The air around you feels damp. You can't wait to get into the air-conditioned store!

Why does the air sometimes feel damp? Moisture enters the air through the **evaporation** of water. Evaporation is the process by which water molecules escape into the air. Through evaporation, the sun's radiant energy turns liquid water into a gas, or water vapor. (The liquid water comes from oceans, rivers, lakes, soil—even from plants and animals.) Winds transport the moisture all over the Earth. At any given time, the atmosphere holds about 14 million tons of moisture! **Water vapor, or moisture, in the air is called humidity.**

The amount of moisture in the air can vary greatly from place to place and from time to time. You will often hear the amount of moisture in the air referred to in terms of **relative humidity** (hyoo-MIHD-uh-tee). Relative humidity is the percentage of moisture the air holds relative to the amount it could hold at a particular temperature.

Suppose that at a certain temperature, 1 kilogram of air can hold 12 grams of water vapor. However, it is actually holding 9 grams. The relative humidity of the air at that temperature would be $9/12 \times 100$, or 75 percent. If the same kilogram of air held 12

Figure 1–21 *Fog, seen here enveloping the skyscrapers of Chicago, is formed when water vapor in the air condenses at a temperature called the dew point. Dew that forms overnight can often be seen on leaves or blades of grass or pine needles in the early morning.*

RELATIVE HUMIDITY					
Dry-Bulb Thermometer Readings (°C)	Difference Between Wet- and Dry-Bulb Thermometer Readings (°C)				
	1	2	3	4	5
10	88	77	66	55	44
11	89	78	67	56	46
12	89	78	68	58	48
13	89	79	69	59	50
14	90	79	70	60	51
15	90	80	71	61	53
16	90	81	71	63	54
17	90	81	72	64	55
18	91	82	73	65	57
19	91	82	74	65	58
20	91	83	74	66	59
21	91	83	75	67	60
22	92	83	76	68	61
23	92	84	76	69	62
24	92	84	77	69	62
25	92	84	77	70	63
26	92	85	78	71	64
27	92	85	78	71	65
28	93	85	78	72	65
29	93	86	79	72	66
30	93	86	79	73	67

Figure 1–22 *If the difference between the readings on the wet-bulb and dry-bulb thermometers is 2°C and the air temperature is 19°C, what is the relative humidity?*

grams of water vapor, it would be holding all the moisture it could hold at that temperature. The relative humidity would then be 100 percent ($12/12 \times 100$). When the relative humidity is 100 percent, the air is said to be saturated. That is, it is holding all the water vapor it can hold at that particular temperature.

Measuring Relative Humidity

Meteorologists measure relative humidity with a **psychrometer** (sigh-KRAHM-uh-ter). A psychrometer consists of two thermometers. The bulb of one thermometer is covered with a moist cloth. This thermometer is the wet-bulb thermometer. The other thermometer is the dry-bulb thermometer.

When air passes over the wet bulb, the water in the cloth evaporates. Evaporation requires heat energy. So evaporation of the water from the cloth cools the thermometer bulb. If the humidity is low, evaporation will take place quickly and the temperature of the wet-bulb thermometer will drop. If the humidity is high, evaporation will take place slowly and the temperature of the wet-bulb thermometer will not change much. In other words, the temperature of the wet-bulb thermometer will be close to the temperature of the dry-bulb thermometer, which is measuring air temperature.

To determine the relative humidity, meteorologists first find the difference between the dry-bulb temperature and the wet-bulb temperature. Then they use a chart similar to the one in Figure 1–22 to find the relative humidity expressed as a percentage. Suppose the dry-bulb thermometer reads 28°C. The difference between the two thermometer readings is 4°C. What is the relative humidity?

Clouds

Warm air can hold more moisture than cold air. As warm, moist air rises in the atmosphere, its temperature begins to drop. Because cold air cannot hold as much water vapor as warm air, the rising air soon becomes saturated. At this point, the water vapor in the air begins to condense, or change into a liquid. The temperature at which water vapor

Figure 1–23 *The three basic cloud types are fluffy cumulus (left), layered stratus (center), and wispy cirrus (right). What type of weather is usually associated with each cloud type?*

condenses is called the dew point. Have you ever seen drops of water, or dew, on blades of grass early in the morning? What do you think caused dew to form on the grass?

Clouds form when moisture in the air condenses on small particles of dust or other solids in the air. The tiny droplets of water that form make up the clouds. A cloud is really a mixture in which particles of a liquid (water) are suspended in a gas (air).

As you can see for yourself just by looking at the sky, clouds come in all sorts of shapes and sizes. Scientists use the basic shape and the altitude of clouds to classify them. The three main types of clouds are cumulus (KYOO-myoo-luhs) clouds, stratus (STRAT-uhs) clouds, and cirrus (SEER-uhs) clouds. Each type of cloud is generally associated with a certain type of weather.

Cumulus clouds look like piles of cotton balls in the sky. These clouds are fluffy and white with flat bottoms. They form at altitudes of 2.4 to 13.5 kilometers. Cumulus clouds usually indicate fair weather. However, when cumulus clouds get larger and darker on the bottom, they produce thunderstorms. These large thunderclouds are called cumulonimbus clouds.

City Showers

The American poet Langston Hughes (1902–1967) was born in Joplin, Missouri, but lived for many years in the section of New York City known as Harlem. His poem "April Rain Song" describes his feelings about a gentle rain shower in the city.

Figure 1–24 *Clouds are classified according to their shape and altitude. Cumulus and stratus clouds that develop at altitudes between 2 and 7 kilometers have the prefix* alto- *in their names. What prefix is used for these clouds if they are at an altitude above 7 kilometers?*

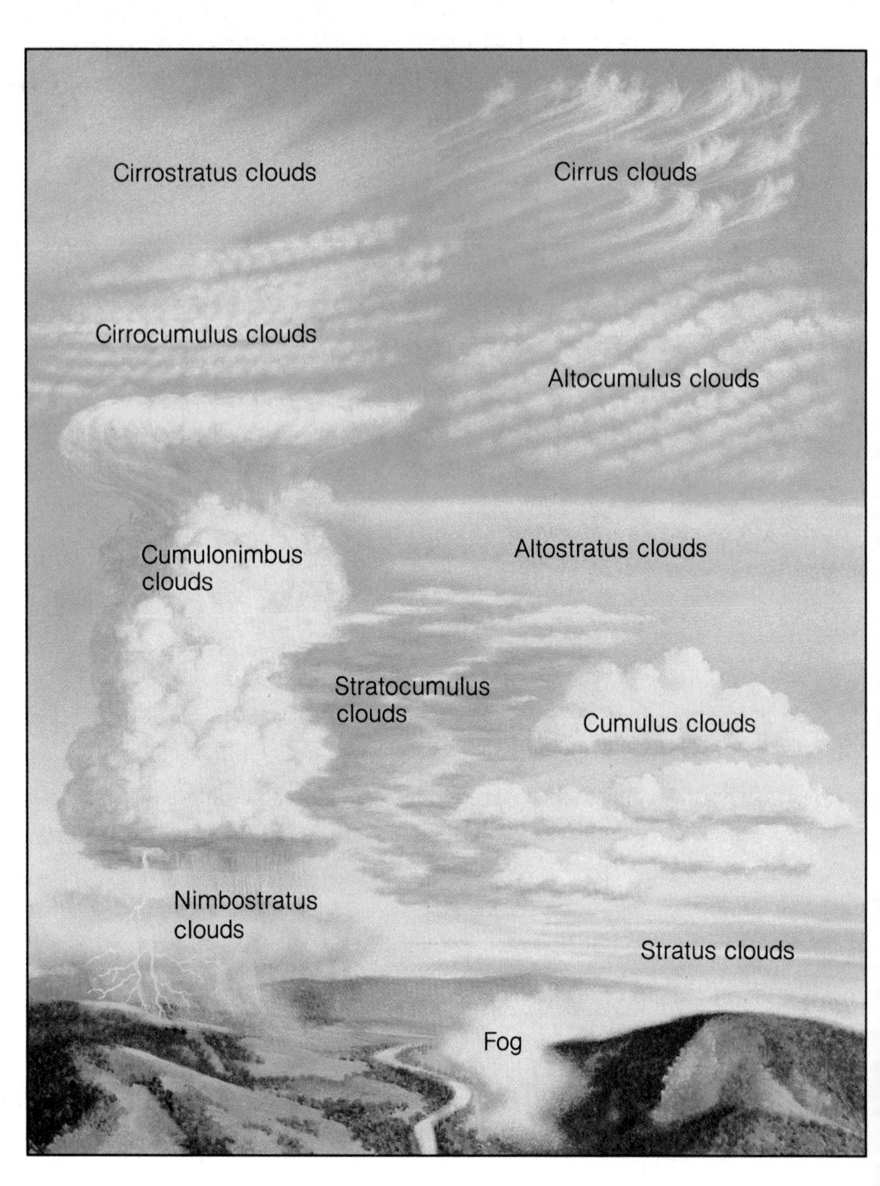

Fog in a Bottle

1. Fill a narrow-necked bottle with hot water. **CAUTION:** *Do not use water hot enough to cause burns.*

2. Pour out most of the water, leaving about 3 centimeters at the bottom of the bottle.

3. Place an ice cube on the mouth of the bottle.

4. Repeat steps 1 through 3 using cold water instead of hot water.

In which bottle does fog form? Why? Why is it helpful to repeat the activity with cold water?

Smooth, gray clouds that cover the whole sky and block out the sun are called stratus clouds. They form at an altitude of about 2.5 kilometers. Light rain and drizzle are usually associated with stratus clouds. Nimbostratus clouds bring rain and snow. When stratus clouds form close to the ground, the result is fog. Ground fog is formed when air above the ground is cooled rapidly at night. Warmer temperatures during the day cause the fog to disappear.

Feathery or fibrous clouds are called cirrus clouds. Sometimes these clouds are called mares' tails. Can you tell from looking at the picture of cirrus clouds in Figure 1–23 why they are called mares' tails? Cirrus clouds form at very high altitudes,

Figure 1–25 *Dark storm clouds gather over a field. What is the name for the type of cloud that usually brings thunderstorms?*

usually between 6 and 12 kilometers. Cirrus clouds are made of ice crystals. You can see cirrus clouds in fair weather, but they often indicate that rain or snow will fall within several hours.

Precipitation

Water vapor that condenses and forms clouds can fall to the Earth as rain, sleet, snow, or hail. Water that falls from the atmosphere to the Earth is called **precipitation** (pree-sihp-uh-TAY-shuhn).

Before water falls to the Earth as precipitation, cloud droplets must increase in size. Cloud droplets increase in size by colliding and combining with other droplets. At some point, the droplets become too large to remain suspended in the cloud. Gravity then pulls these larger drops of water to the Earth as rain. An average raindrop contains about one million times as much water as a cloud droplet!

When falling raindrops pass through an extremely cold layer of air, they sometimes freeze into small ice pellets called sleet. Sleet reaches the Earth only in the winter. Why? What do you think happens to sleet in the summer?

Snow forms when water vapor changes directly into a solid. Snowflakes are flat six-sided ice crystals that have beautiful shapes. Because snowflakes sometimes clump together, it is often hard to see the

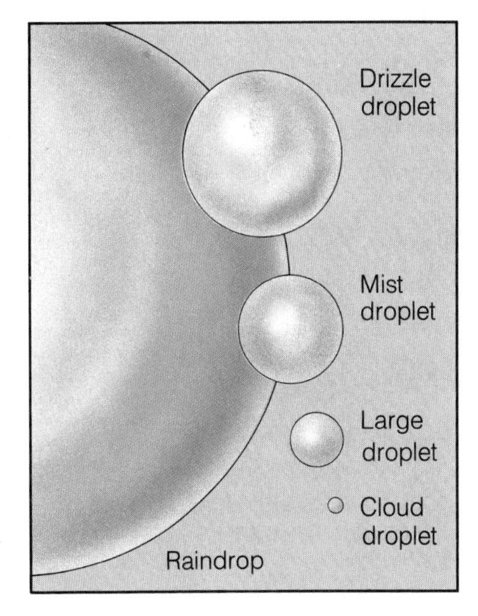

Figure 1–26 *Notice the relative sizes of water droplets in a cloud. Which droplet is the smallest? The largest? When water droplets get large enough, they fall to Earth.*

Figure 1–27 *When huge numbers of individual snowflakes (top) accumulate, they form a lovely winter scene (center). These hailstones became large and heavy enough to fall to the ground (bottom).*

separate crystals. If you could look at many individual snowflakes, you would find that they all have different shapes. In fact, no two snowflakes are exactly alike!

Hail is one of the most damaging forms of precipitation. It is usually formed in cumulonimbus clouds. Hailstones are small balls or chunks of ice ranging from 5 to 75 millimeters in diameter. Hailstones are formed when water droplets hit ice pellets in a cloud and freeze. If the updraft (upward movement of the wind) is strong enough, the hailstones remain in the cloud for a long time. As more water droplets strike them, new layers of ice are added. Finally, the hailstones get so big and heavy that they fall to the ground. One of the largest hailstones ever found fell on Coffeyville, Kansas, in September 1970. This hailstone measured 140 millimeters in diameter!

Measuring Rainfall

Precipitation in the form of rain is measured with a **rain gauge.** A rain gauge is a straight-sided container with a flat bottom that collects rain as it falls. The amount of rain collected in a rain gauge over a given period of time is usually expressed in millimeters or centimeters. Weather observers in the United States, however, usually express the amount of rainfall in inches.

1–4 Section Review

1. What is the difference between humidity and relative humidity?
2. How does moisture enter the air?
3. What are the three main types of clouds? What are the four main types of precipitation?
4. How is relative humidity measured? How is rainfall measured?

Critical Thinking—*Making Calculations*

5. Suppose 1 kilogram of air can hold 16 grams of water vapor but actually holds 8 grams. What is the relative humidity?

1–5 Weather Patterns

Guide for Reading

Focus on these questions as you read.

- *What are the four major air masses that affect the weather in the United States?*
- *How do different types of fronts and storms affect weather in the United States?*

As you have already learned in the first part of this chapter, there are many factors that interact in the Earth's atmosphere to cause weather. You know from experience that weather is constantly changing. Clouds and rain may move quickly into an area and move away just as quickly. Days of blue, sunny skies and warm temperatures may change overnight to gray, stormy skies and freezing rain. What atmospheric conditions and patterns cause these changes in the weather?

Air Masses

Changes in the weather are caused by movements of large bodies of air called **air masses.** Air masses usually cover thousands of square kilometers. The properties of the air in an air mass are nearly the same, or uniform, throughout the air mass. Like clouds, air masses are classified according to some basic characteristic. For clouds, the characteristic is shape; for air masses, it is where they form. Where air masses form determines two of their most important properties: temperature and humidity, or moisture content. Air masses that form over tropical regions are warm. Those that form over polar regions are cold. Continental air masses, which form over continents, are relatively dry. Maritime air masses form over oceans, and they are relatively humid. **The four major types of air masses that affect the weather in the United States are maritime tropical, maritime polar, continental tropical, and continental polar.** Figure 1–28 on page 36 shows where these air masses form and the areas of the United States they most influence.

The maritime tropical air mass forms over the ocean near the equator. It holds warm, moist air. In the summer, the maritime tropical air mass brings hot, humid weather. But if the warm, moist air comes in contact with a cold air mass in the winter, rain or snow will fall.

The maritime polar air mass forms over the Pacific Ocean in both winter and summer. It forms over the cold North Atlantic waters in the summer.

Does Air Have Mass?

1. Using a balance, find the mass of a fully deflated ball.

2. Use an air pump to completely inflate the ball. What do you think the mass of the inflated ball will be?

3. Place the inflated ball on the balance and find its mass. What is the mass of the inflated ball? Was your prediction correct?

■ Does air have mass? How do you know?

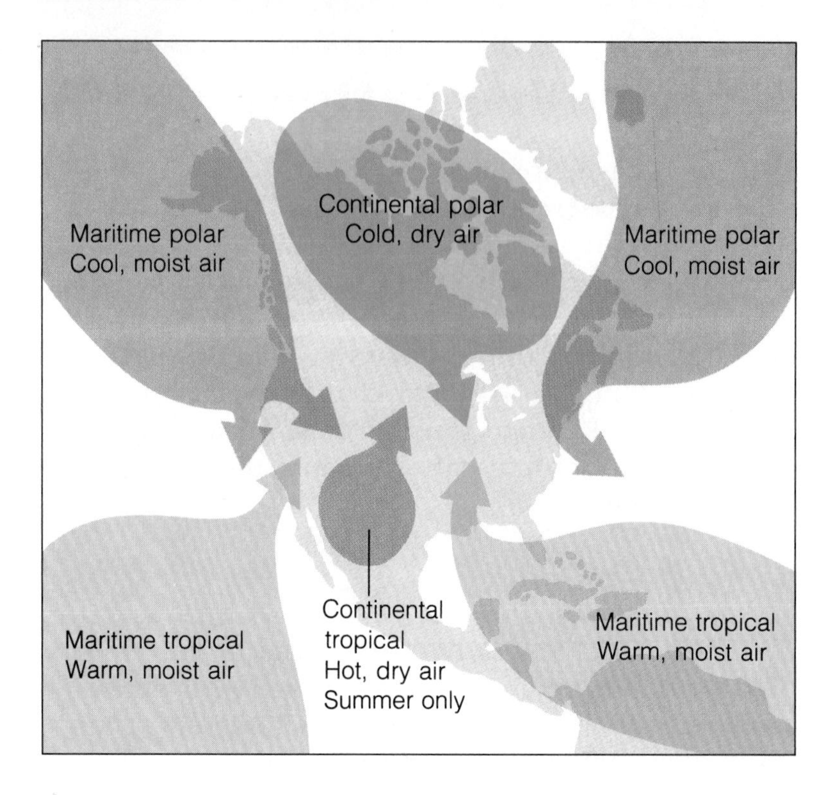

Figure 1–28 *Four major air masses affect the weather of the United States. Each air mass brings different weather conditions. Which air masses affect the weather where you live?*

During the summer, the maritime polar air mass brings cooler temperatures to the eastern states and fog to California and other western states. Heavy snow and cold temperatures are produced by the maritime polar air mass in the winter.

During the summer, the continental tropical air mass forms over land in Mexico. This air mass brings dry, hot air to the southwestern states. The continental polar air mass forms over land in northern Canada. In winter, this cold, dry air mass causes extremely cold temperatures in the United States.

Fronts

When two air masses that have different properties meet, they do not mix readily. Instead, a boundary forms between the two air masses. This boundary is called a **front.** (The term front was first applied to weather during World War I, when opposing armies faced one another across a battlefront.) The weather at a front is usually unsettled and stormy. **There are four different types of fronts: cold fronts, warm fronts, occluded fronts, and stationary fronts.**

A cold front forms when a mass of cold air meets and replaces a mass of warm air, as shown in Figure

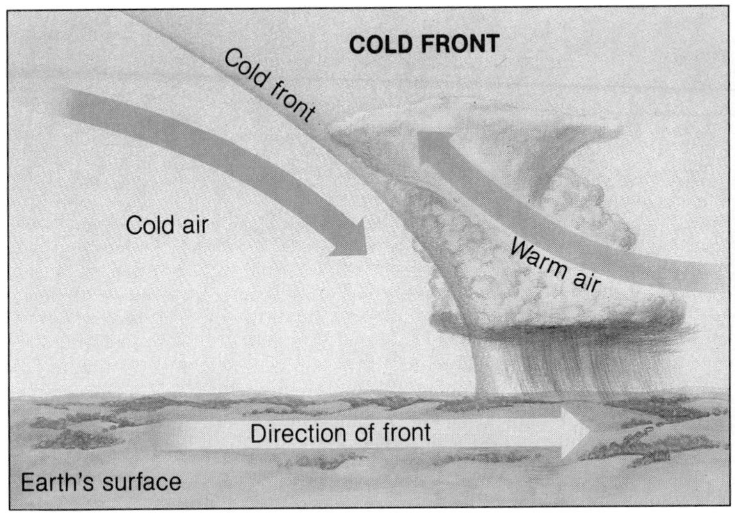

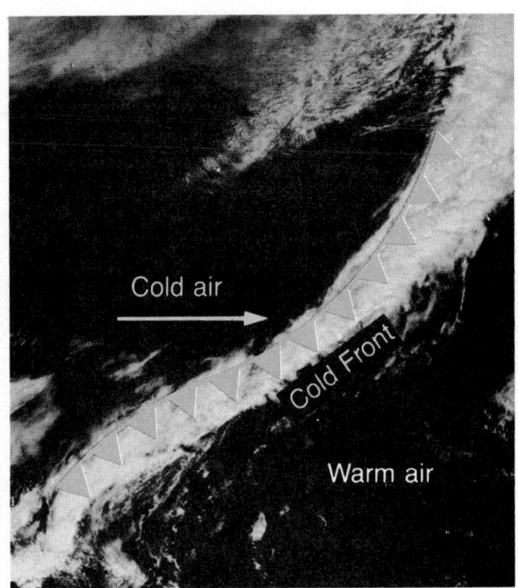

Figure 1–29 *A cold front forms when a mass of cold air meets and replaces a mass of warm air, as shown in the illustration (left). Notice the weather symbols showing the edge of the cold front on the satellite weather photograph (right).*

1–29. The cold air mass forces its way underneath the warm air mass and pushes it upward. What do you know about cold air and warm air that explains why this happens? Violent storms are associated with a cold front. Fair, cool weather usually follows.

A warm front forms when a mass of warm air overtakes a cold air mass and moves over it. This process is shown in Figure 1–30. Rain and showers usually accompany a warm front. Hot, humid weather usually follows.

A cold front travels faster than a warm front. When a cold front overtakes a warm front, an occluded front forms. An occluded front is shown in Figure 1–31 on page 38. As the warm air is pushed upward, the cold air meets cool air. An occluded

Figure 1–30 *A warm front forms when a mass of warm air overtakes a cold air mass and moves over it, as shown in the illustration (left). Notice the weather symbols showing the edge of the warm front on the satellite weather photograph (right).*

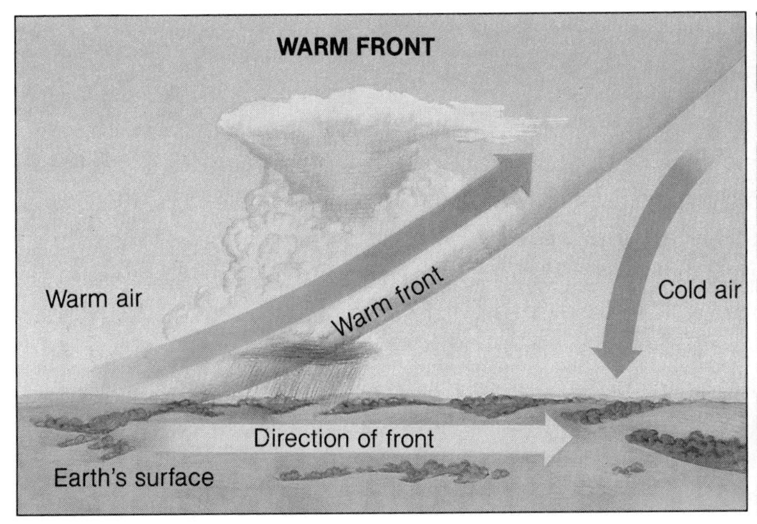

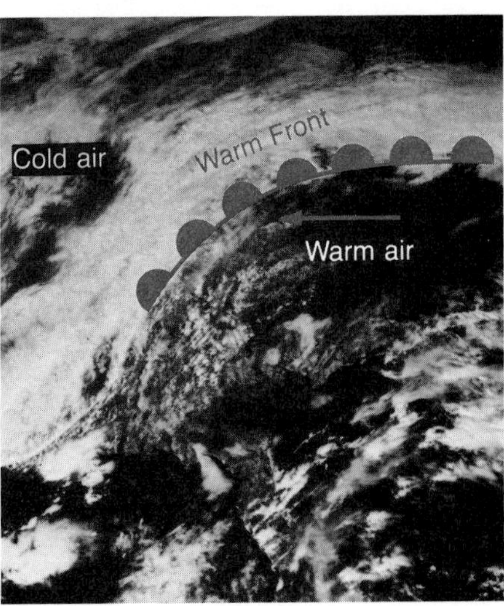

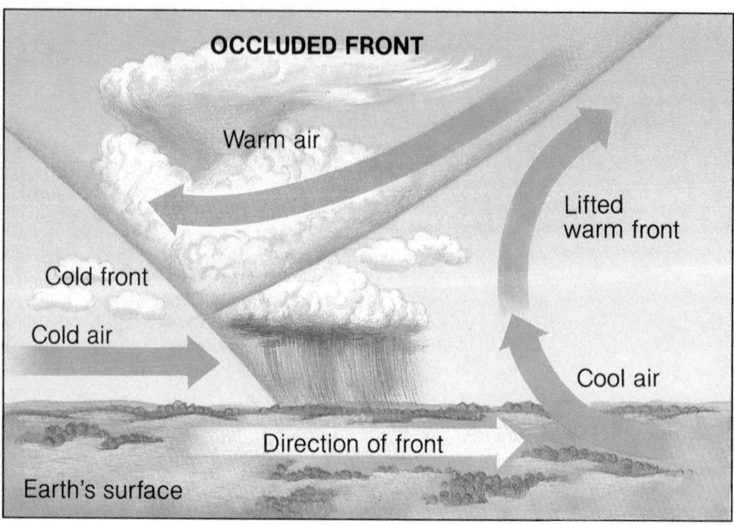

Figure 1–31 *An occluded front forms when a cold front overtakes a warm front, as shown in the illustration (right). Notice the weather symbols showing the occluded front on the satellite weather photograph (left).*

front may also occur when cool air overtakes a cold front and warm air is pushed upward. An occluded front produces less extreme weather than a cold front or a warm front.

When a warm air mass meets a cold air mass and no movement occurs, a stationary front forms. Rain may fall in an area for many days when a stationary front is in place.

Storms

A storm is a violent disturbance in the atmosphere. It is marked by sudden changes in air pressure and rapid air movements. Some storms may cover a huge area, whereas others cover only a small area. You are probably familiar with one or more of the storms described here. As you read about these storms, compare the descriptions with your own experiences.

RAINSTORMS AND SNOWSTORMS When two different fronts collide, rainstorms or snowstorms form. When a warm front moves in and meets a cold front, heavy nimbostratus clouds develop. In the summer, the result is a steady rainfall that lasts for several hours. In the winter, a heavy snowfall occurs. If the wind speed is more than 56 kilometers per hour and the temperature is below –7°C, a blizzard results.

Sometimes, heavy rain falling over a wide area freezes instantly on trees, power lines, and other surfaces. The result is an ice storm. Although the layer

Figure 1–32 *An ice storm covers everything in its path like frosting on a cake or glaze on a doughnut. Although it may look beautiful, an ice storm can cause much damage.*

of glittering ice may look beautiful, ice storms can cause great damage by knocking down trees and power lines. Some interesting effects of an ice storm are shown in Figure 1–32.

THUNDERSTORMS When a cold front moves in and meets a warm front, cumulonimbus clouds produce thunderstorms. Thunderstorms are heavy rainstorms accompanied by thunder and lightning. These storms can be quite dangerous. Violent downdrafts and strong wind shear (a great change in wind velocity over a short distance) are often associated with thunderstorms. These conditions are of great concern to airplane pilots and air-traffic controllers during take-offs and landings.

The other factor that makes thunderstorms dangerous is lightning. What is lightning? You may have heard the story about Benjamin Franklin flying a kite into a thunderstorm to prove that lightning is a form of electricity. During a thunderstorm, areas of positive and negative electric charges build up in the storm clouds. Lightning is a sudden discharge, or spark, of electricity between two clouds or between a cloud and the ground.

Lightning striking the ground is the leading cause of forest fires in the western states. Lightning may also strike people, animals, or buildings. In fact, more people are killed every year by lightning than as a result of any other violent storm! There are some important safety rules you should remember when you see a lightning storm coming. Avoid open

Activity Bank

What Causes Lightning?, p.122

ACTIVITY
CALCULATING

Storms and Lightning

Every day, an average of 45,000 storms occur across the Earth's surface. How many storms occur in a year?

Lightning can occur with or without a thunderstorm. Lightning strikes somewhere in the United States about 50 million times a year. On the average, how many times each day does lightning strike the United States?

Figure 1–33 *Why should you never stand under a tree during a thunderstorm?*

Speed of Sound

1. Working with a partner, measure a distance of 100 meters in an open outdoor area.

2. Stand at one end of this 100-meter distance with a stopwatch. Have your partner stand at the opposite end and make a loud, sharp noise—for example, by striking an old aluminum pot with a metal spoon.

3. Start the stopwatch exactly when you see your partner strike the pot. Stop the watch exactly when you hear the noise. How long did it take for you to hear the noise?

4. Repeat the procedure two more times and calculate the average for the three trials. Divide the distance by the average time to find the speed of sound in meters per second.

spaces, but do not take shelter under a tree. The best shelter is inside a building. Remember, however, to stay away from sinks, bathtubs, and televisions. (Some people have even been injured by lightning when talking on the telephone.) And never try to repeat Benjamin Franklin's experiment with a kite!

Loud thunder claps usually accompany the lightning in a thunderstorm. The electrical discharge of lightning heats the air. When the air is heated, it expands rapidly. This sudden expansion of the air results in sound waves, which we hear as thunder. Do you know why you see lightning before you hear thunder? Although lightning and thunder occur at the same time, you see the lightning almost instantly. The sound waves of thunder, however, travel much more slowly than light (340 meters per second compared with 300,000 kilometers per second). If you hear thunder about 3 seconds after you see a flash of lightning, the lightning is about 1 kilometer away. How can you tell if a thunderstorm is moving toward or away from you?

CYCLONES AND ANTICYCLONES Air pressure has a great effect on the weather. An area of low pressure that contains rising warm air is called a cyclone. In a cyclone, cooler air moves in to take the place of the rising warm air. The air currents begin to spin. Winds spiral around and into the center of the cyclone. The winds move in a counterclockwise direction in the Northern Hemisphere. Cyclones usually cause rainy, stormy weather. What do you think causes the air currents in a cyclone to spin?

A high-pressure area that contains cold, dry air is called an anticyclone. Winds spiral around and out from the center of an anticyclone. In the Northern Hemisphere, the winds move in a clockwise direction. The weather caused by anticyclones is usually clear, dry, and fair.

HURRICANES A hurricane is a powerful cyclone (a low-pressure area containing rising warm air) that forms over tropical oceans. Hurricanes that form over the western Pacific Ocean are called typhoons. During late summer and early autumn, low-pressure areas often form over the Caribbean or the Gulf of Mexico. Warm, moist air begins to rise rapidly. Cooler air moves in, and the air begins to spin. As

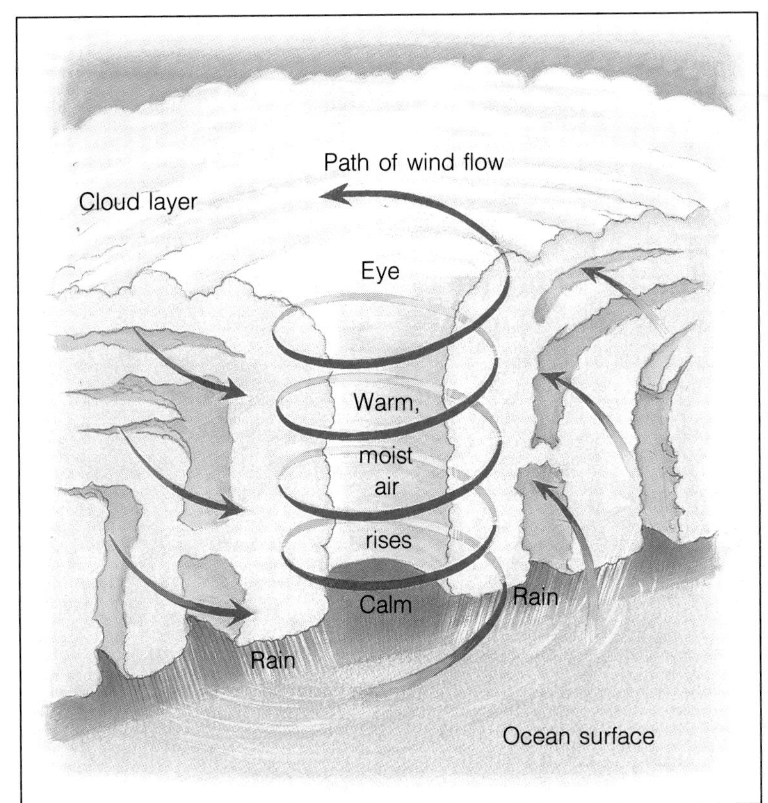

Figure 1-34 *This photo shows some of the destruction caused by Hurricane Andrew when it hit south Florida in August 1992. A hurricane has a large counterclockwise movement of air surrounding a low-pressure center. What is the name for the calm center of a hurricane?*

the air pressure in the center drops, more air is drawn into the spinning system. The system begins to spin faster. The rapidly spinning, rising air forms a doughnut-shaped wall of strong winds, clouds, and rainfall. Inside the wall, the air is calm. This calm center is called the eye of the hurricane. Outside the eye, winds may reach speeds between 120 and 320 kilometers per hour.

The high waves and strong winds of a hurricane often cause great damage, especially in coastal areas. Heavy rain may also cause serious flooding. Meteorologists can track the path of a hurricane and issue watches or warnings to people living near the coast as the storm approaches. A typical hurricane lasts for about 9 days. In extreme cases, hurricanes can last as long as 3 to 4 weeks. In terms of the total energy involved, hurricanes are the most powerful storms on Earth. As a hurricane moves inland, it loses its force and power. Can you explain why?

TORNADOES Tornadoes are also incredibly destructive. A tornado is a whirling, funnel-shaped cloud. It develops in low, heavy cumulonimbus clouds. The air pressure at the bottom of the funnel

Figure 1–35 *A tornado lasts no more than a few minutes, but it is extremely destructive. You can see just how destructive from this photograph of the aftermath of a tornado in Raleigh, North Carolina.*

of swirling air is extremely low. When this low-pressure area touches the ground, it acts like a giant vacuum cleaner. Some tornadoes occur over water. A tornado over a lake or ocean is called a waterspout.

Meteorologists are not sure how tornadoes form. Tornadoes occur most often in spring during the late afternoon or early evening. In the United States, they are most common on the Great Plains. In fact, tornadoes are so common that this part of the United States is often called Tornado Alley. Here, cool, dry air from the west collides with warm, moist air from the Gulf of Mexico.

The diameter of an average tornado is only about 0.4 kilometer. The length of a tornado's path varies, but it averages 6 kilometers. Tornadoes generally last only a few minutes. But because they are so concentrated, they are intensely violent and dangerous storms. Tornadoes have strong winds that can reach speeds of more than 350 kilometers per hour. Roofs and walls of buildings may be blown out by the winds. Houses, railroad cars, automobiles, and even people may be picked up and thrown hundreds of meters. A tornado in Nebraska tossed a 225-kilogram baby grand piano almost 400 meters across a corn field!

1–5 Section Review

1. What are two important properties of an air mass? What are the four major types of air masses that affect the weather in the United States?
2. What is a front? Name four different types of fronts.
3. What is the difference between a cyclone and an anticyclone?
4. Describe the differences between a hurricane and a tornado.

Connection—*Geography*

5. Saskatchewan, Canada, has hot summers, cold winters, and stormy weather in spring and autumn. Explain this weather in terms of major air masses.

1–6 Predicting the Weather

Guide for Reading

Focus on this question as you read.

- *How do meteorologists interpret weather data to make weather forecasts?*

As Lisa was about to leave for school, her mother called after her, "Don't forget to take your umbrella." Lisa looked out the window. The sun was shining brightly, and there were only a few fluffy white clouds in the sky. How could Lisa's mother know that she would need her umbrella later in the day?

Accurate weather predictions are important for planning many human activities. Farmers need to know the best times to plant and harvest their crops. Airplane takeoffs, landings, and flight paths are scheduled according to local weather conditions. Weather forecasts alert people to severe storms that could endanger life or property. And most people want to know what the weather will be like as they go to and from work and school or plan outdoor activities.

Meteorologists rely on up-to-date observations of current weather conditions to make their forecasts. Most weather forecasts are made for periods of a few hours up to five days. **Meteorologists interpret weather information from local weather observers, balloons, satellites, and weather stations around the world.** Weather stations are located on land and on ships at sea. Meteorologists also use computers to help them interpret weather data.

Figure 1–36 *Here you see two meteorologists releasing a weather balloon. A computer monitor displays weather data transmitted from a weather satellite.*

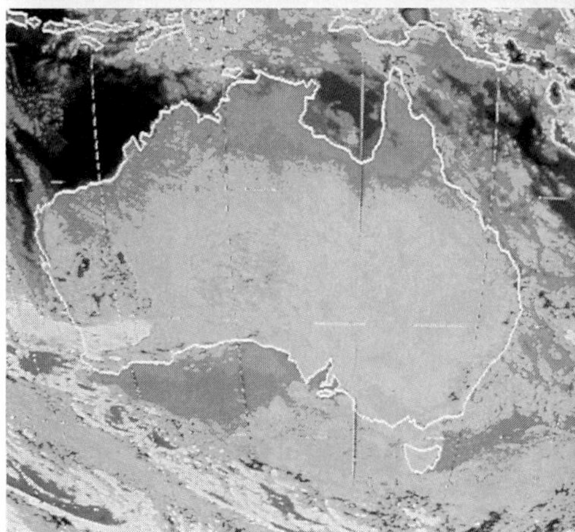

Weather Maps

Accurate weather forecasts are made possible by studying information about atmospheric conditions at several places. In the United States, weather data are gathered from more than 300 local weather stations. These data include temperature, air pressure, precipitation, and wind speed and direction. Weather data are used to prepare a daily weather map. Information about cloud cover, air masses, and fronts may also be included on a weather map. So a weather map is a "picture" of local weather conditions.

The information on weather maps is often recorded in the form of numbers and symbols. Symbols are used to show wind speed and direction,

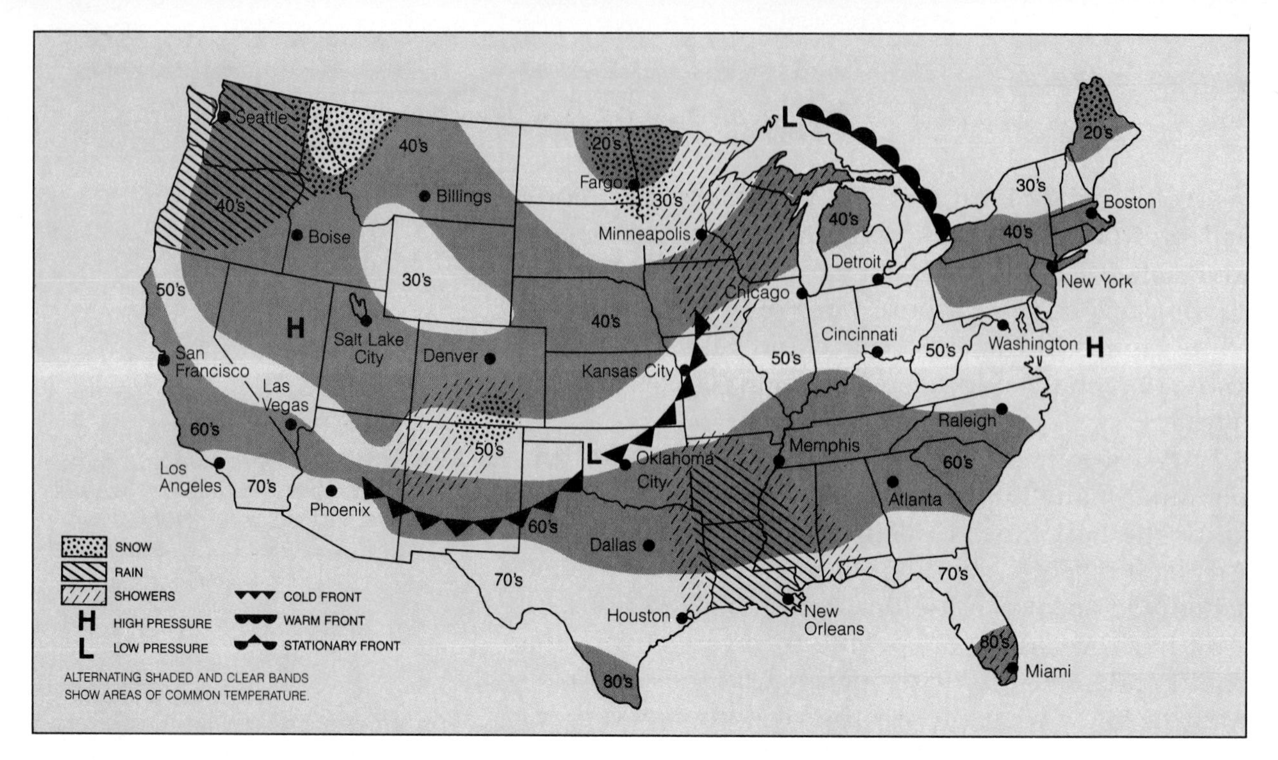

Figure 1–37 *This weather map is similar to those often printed in newspapers. The legend at the bottom left tells you how to read the symbols on this map. These symbols are not exactly the same as those on an official National Weather Service map. What was the weather in Minneapolis on this particular day?*

cloud cover, precipitation, position and direction of fronts, and areas of high and low pressure. The symbols on official National Weather Service weather maps are used by all nations. As with other branches of science, it is important that meteorologists from different countries be able to understand universal weather symbols. In chemistry, universal symbols are used to identify the different chemical elements. For example, chemists in the United States, Russia, Japan, and China all recognize H as the symbol for hydrogen and C as the symbol for carbon. In the same way, meteorologists all over the world recognize the universal weather symbols. You might notice, however, that the official National Weather Service symbols differ from those used on the simplified weather map in your local newspaper. A typical newspaper weather map is shown in Figure 1–37.

Recording Weather Data

Look closely at Figure 1–38, which shows how weather data from a particular location are presented. The circle represents an observation station. Weather data are placed in specific positions inside and outside the station circle. Think of the station circle as the point of an arrow. Attached to the station circle is a line, which is like the shaft of an arrow. The wind direction is represented as moving along the shaft of the arrow toward the station circle. The wind direction is the direction from which the wind is blowing. According to this station circle, how is the wind blowing? You are correct if you said from southwest to northeast.

Small lines at the end of the shaft are symbols for the wind speed. Each full line represents a wind speed of 9–14 miles per hour. (Notice that in the United States, English units are commonly used instead of metric units.) In this station circle, the wind speed is about 18 miles per hour. Average daily temperature is given in degrees Fahrenheit next to the station circle. Other data shown include percentage of cloud cover and atmospheric pressure in millibars and inches of mercury. One millibar equals about 0.03 inch of mercury. (Inches of mercury are usually given to the nearest hundredth and range from 28.00 to 31.00 inches.) What were the temperature and percentage of cloud cover when the data at this weather station were recorded?

The data from weather stations all around the country are assembled into weather maps at the National Weather Service. Figure 1–39 on page 46 shows such a weather map. Notice that this map includes most of the weather station data shown in Figure 1–38. How does this weather map compare with weather maps in your local newspaper?

A warm front is shown on a weather map as a line with half circles pointing in the direction of its movement. A cold front is shown as a line with triangles pointing in the direction of its movement. To show a stationary front, the symbols for a warm front and a cold front are combined. Why is this appropriate? The symbols are shown as pushing against each other, to illustrate that a stationary front does not move in either direction. The symbols for a warm front and a cold front are also combined to show an occluded front. But the symbols are on the same side to illustrate that both fronts are moving in the same direction.

Isotherms and Isobars

On some weather maps, you may see curved lines called **isotherms** (IGH-soh-thermz). The word

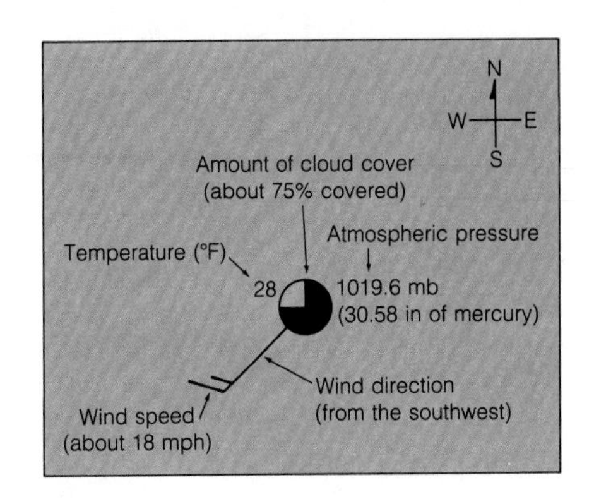

Figure 1–38 *These weather data indicate readings taken at a weather observation station.*

CAREERS

Weather Observer

Accurate weather reports depend on **weather observers** located at observation stations all around the country. At the same time every day, four times a day, these observers collect weather data, such as wind speed, wind direction, air pressure, and relative humidity.

If you would like to learn more about weather observers, write to the United States Department of Commerce, NOAA, National Weather Service, Silver Spring, MD 20910.

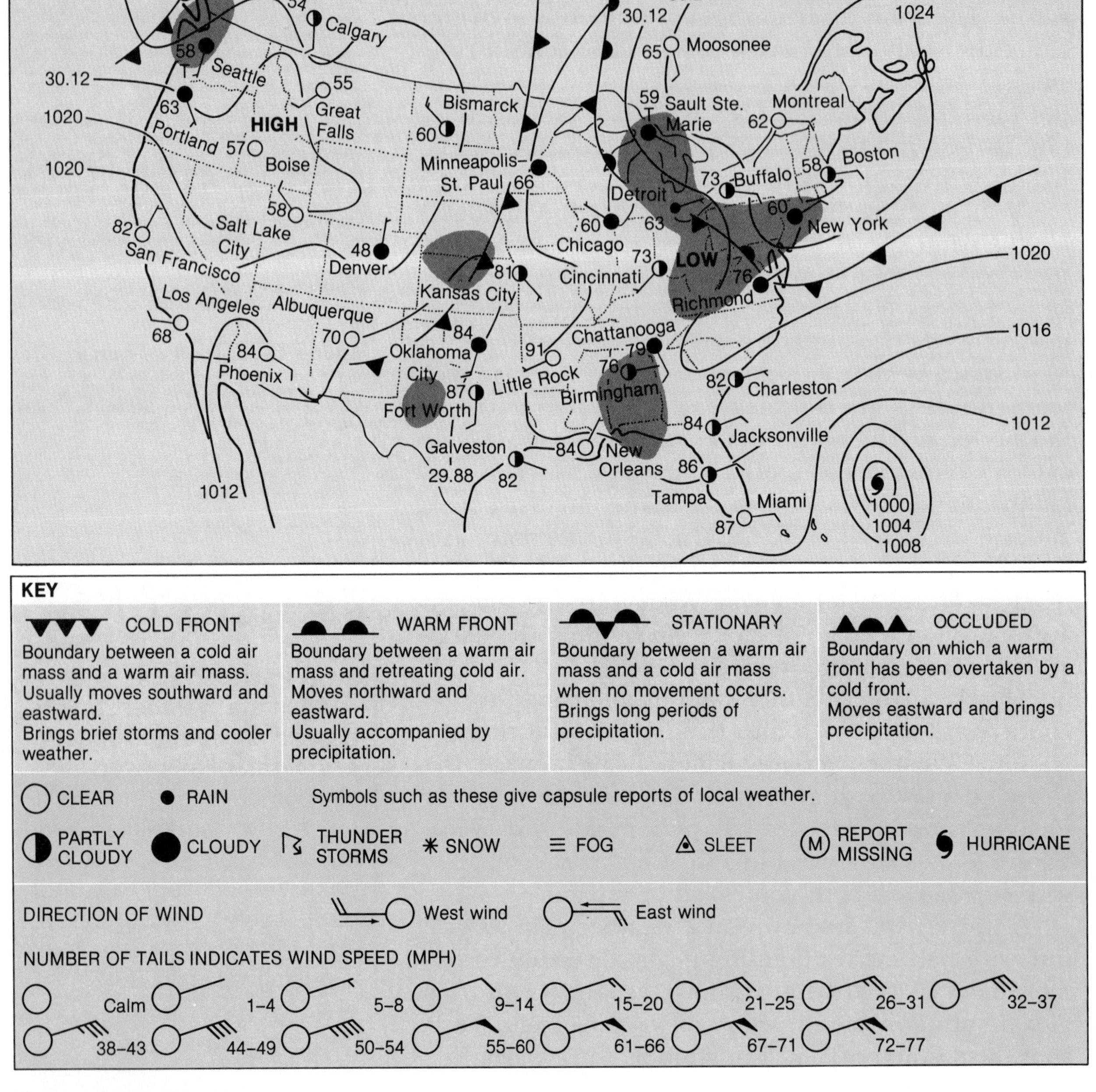

Figure 1–39 *This illustration shows a typical weather map with data from observation stations all over the country. Use the key below the map to determine the weather conditions in your state on this particular day.*

isotherm is made up of the prefix *iso-*, meaning equal, and the root word *-therm,* meaning heat. Isotherms are lines that connect locations with the same temperature. The number on the end of an isotherm indicates the temperature at all points along the isotherm.

Notice the curved lines running through the weather map in Figure 1–39. These lines are called **isobars** (IGH-soh-barz). From the prefix and root word, what do you think isobars are? Isobars are lines that join places on a weather map that have the same air pressure. The number at the end of an isobar represents the air pressure recorded at each observation station along the isobar. The air pressure can be given in millibars, inches of mercury, or both. On this weather map, the isobars are marked at one end with air pressure in inches and at the other end with air pressure in millibars.

Controlling the Weather

An American writer and humorist once said, "Everybody talks about the weather, but nobody does anything about it." Actually, people have always tried to do something about the weather. Because rain is so important for the growth of crops, many efforts to control the weather have centered around rainmaking. Native Americans tried to encourage rainfall by performing elaborate rain dances. In 1901, French farmers fired an "antihail" cannon into the air. They were hoping to break up large hailstones that could destroy their crops and to produce a gentle rain instead. But all they got was a loud bang!

If something could be done about the weather, the results would be important to many people. By controlling the weather, damage from hailstorms, tornadoes, lightning, and hurricanes could be avoided. Droughts and floods could be prevented by controlling rainfall.

At the present time, weather control is limited to the seeding of clouds. Cloud seeding involves the sprinkling of dry ice (solid carbon dioxide) or silver iodide crystals into supercooled layers of stratus clouds. (Supercooling occurs when water remains a liquid below its freezing point.) Seeding causes water droplets to evaporate, or change into a gas. As they evaporate, the water droplets absorb heat from nearby supercooled droplets. The supercooled droplets then freeze and form ice crystals. The crystals grow rapidly. At some point, the crystals become large enough to fall to the Earth as rain or snow.

Experiments have shown that seeding hailstorms and hurricanes decreases their force. However, the most successful use of cloud seeding has been in the partial removal of cool fog at airports. Dry ice is sprinkled onto the fog, causing ice crystals to form. As a result, the fog loses some of its moisture. In

Figure 1–40 *A Sioux member of the Big Thunder Dance Company is shown performing a rain dance.*

ACTIVITY DOING

Observing and Predicting the Weather

1. Cut out the weather map from your local newspaper every day for a week. Each day note what type of weather is predicted for your area for the following day.

2. Each day write a brief description of the actual weather conditions.

3. Obtain a copy of the *Farmer's Almanac* and read the weather predictions for those days.

How do the long-range predictions in the *Farmer's Almanac* compare with the daily weather map predictions and your observations?

Figure 1–41 *Notice how this airplane is releasing silver iodide crystals from the back of the wing as it flies through the clouds. Cloud seeding was first used in the 1940s as a method of producing rain during dry periods. Earlier in the century, some people were willing to try more creative methods to cause rainfall!*

this way, a clear area, or "hole," is made in the fog so airplanes can take off and land. Unfortunately, most fog is warm and is not affected by seeding. But warm fog can be removed by mixing the fog with warmer, drier air from above or by heating the air from the ground. The fog evaporates when it is heated.

In the future, we may be able to improve living conditions on the Earth by controlling the weather. Do you think weather control could damage the environment? If so, in what way?

1–6 Section Review

1. What are some sources of weather information used by meteorologists?
2. What are the lines on a weather map that connect areas of equal pressure? Equal temperature?
3. Using the weather map in Figure 1–39, describe the weather conditions in Fort Worth. What is the wind speed in Denver? What is the temperature in San Francisco?
4. Why is controlling the weather important?

Critical Thinking—*Making Predictions*

5. Predict how your life might change if the weather could be carefully controlled.

CONNECTIONS

Desalting California's Water Supply

Schoolchildren, hoping for sunny weather, chant, "Rain, rain go away. Come again another day." But what happens if the rain goes away and does *not* come again? Too little precipitation in an area can result in a drought—a prolonged period of dry weather. The United States experienced the worst drought in its history during the 1930s.

In 1991, California was in the fifth year of a drought that was the worst since the great drought of the 1930s. Animals, plants, and people suffered from the lack of water. With reservoirs down to half their normal levels, many cities had to look elsewhere for their water supplies. So they turned to the Pacific Ocean. Cities such as Santa Barbara made plans to build desalting plants to turn seawater into fresh water.

Desalting plants have been used for more than 30 years in very dry parts of the world such as the Middle East. These desalting plants remove the salts and other impurities from seawater and produce fresh water. The largest desalting plant in the world, in Saudi Arabia, produces more than 1 billion liters of fresh water a day from the salty water of the Persian Gulf.

There are two basic kinds of desalting technology. Most desalting plants use a distillation process in which the seawater is boiled to produce steam. The salt-free steam is then condensed (changed to a liquid), and the resulting fresh water is collected and stored. Another kind of desalting process is called reverse osmosis. In this process, seawater is filtered to remove suspended and dissolved solids.

Unfortunately, it is expensive to make fresh water from seawater—as much as four times as expensive as obtaining fresh water from natural sources. However, as natural sources of fresh water in California become less available and more costly, desalting may become a more practical alternative.

Dried-up reservoir in California (top)

Desalting plant in Saudi Arabia (bottom)

Laboratory Investigation

Using a Sling Psychrometer to Determine Relative Humidity

Problem

How can you determine relative humidity using a handmade sling psychrometer?

Materials *(per group)*

sling psychrometer
medicine dropper

Procedure

1. Using a medicine dropper, add a few drops of water to the gauze on one thermometer. This is the wet-bulb thermometer.
2. Hold the dowel in your hand and slowly spin the thermometers around the nail. This spinning motion will speed up the evaporation process. **Note:** *Be sure to stand away from other students and to spin the thermometers slowly.*
3. From time to time, check the temperature of the wet-bulb thermometer. Keep spinning the thermometers until the temperature stops dropping.
4. When the temperature of the wet-bulb thermometer has stopped dropping, read the temperatures on both the wet-bulb and dry-bulb thermometers. Calculate the difference between the two readings.
5. Using the dry-bulb temperature and the difference between the dry-bulb and the wet-bulb temperatures, determine the relative humidity. (Refer to Figure 1–22.) Express your answer as a percentage.

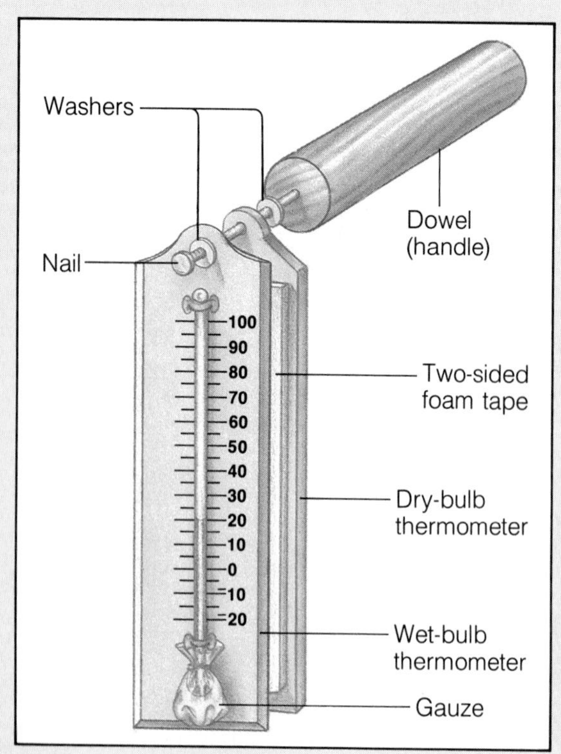

Observations

1. Which of the two thermometers measures the air temperature?
2. What is the relative humidity in your classroom?

Analysis and Conclusions

1. What is the relationship between evaporation and the wet-bulb temperature?
2. What is the relationship between evaporation and relative humidity?
3. What would the relative humidity be if both the wet-bulb thermometer and the dry-bulb thermometer measured the same temperature? Explain.
4. **On Your Own** How do you think the relative humidity inside your classroom compares with the relative humidity outdoors? How could you find out whether you are right or wrong?

Study Guide

Summarizing Key Concepts

1–1 Heating the Earth

▲ Factors that interact to cause weather are heat energy, air pressure, winds, and moisture in the air.

▲ Heat energy is transferred by conduction, convection, or radiation.

▲ Air temperature varies depending on the angle at which the sun's rays strike the Earth.

1–2 Air Pressure

▲ Air pressure depends on the density of the air.

▲ Factors affecting air pressure are temperature, water vapor in the air, and elevation.

1–3 Winds

▲ Local and global winds are caused by differences in air pressure due to unequal heating of the air.

▲ Local winds blow from any direction and cover short distances; global winds blow from a specific direction and cover long distances.

1–4 Moisture in the Air

▲ Water vapor, or moisture, in the air is called humidity.

▲ There are three main types of clouds: cumulus, stratus, and cirrus.

▲ Water vapor that condenses and forms clouds can fall to the Earth as precipitation in the form of rain, sleet, snow, or hail.

1–5 Weather Patterns

▲ The four major air masses that affect weather in the United States are maritime tropical, maritime polar, continental tropical, and continental polar.

▲ When two air masses meet, a cold front, a warm front, an occluded front, or a stationary front may form.

1–6 Predicting the Weather

▲ Meteorologists use data from local weather observers, balloons, satellites, and weather stations to predict the weather.

Reviewing Key Terms

1–1 Heating the Earth

atmosphere
conduction
convection
radiation
greenhouse effect
thermometer

1–2 Air Pressure

air pressure
barometer

1–3 Winds

wind
sea breeze
land breeze
Coriolis effect
anemometer

1–4 Moisture in the Air

evaporation
relative humidity
psychrometer
precipitation
rain gauge

1–5 Weather Patterns

air mass
front

1–6 Predicting the Weather

isotherm
isobar

Chapter Review

Content Review

Multiple Choice

Choose the letter of the answer that best completes each statement.

1. The factors that cause weather include heat, air pressure, wind, and
   a. temperature. c. moisture.
   b. elevation. d. storms.
2. Most of the heat energy in the Earth's atmosphere is transferred by
   a. conduction. c. radiation.
   b. convection. d. ultraviolet rays.
3. Isobars on a weather map connect places with the same
   a. temperature. c. precipitation.
   b. wind speed. d. air pressure.
4. A wind that blows from the sea to the land is called a
   a. land breeze. c. jet stream.
   b. trade wind. d. sea breeze.
5. A powerful storm that forms over tropical oceans is called a
   a. thunderstorm. c. hurricane.
   b. tornado. d. blizzard.
6. Air pressure is measured with a
   a. thermometer. c. anemometer.
   b. barometer. d. psychrometer.
7. The type of front formed when a mass of warm air moves over a mass of cold air is a(an)
   a. cold front. c. occluded front.
   b. warm front. d. stationary front.
8. Fair weather clouds that may develop into thunderstorms are called
   a. cumulus clouds. c. cirrus clouds.
   b. stratus clouds. d. nimbostratus clouds.

True or False

If the statement is true, write "true." If it is false, change the underlined word or words to make the statement true.

1. Winds that blow from a specific direction are called local winds.
2. Ozone gas in the Earth's atmosphere absorbs infrared rays from the sun.
3. Isotherms on a weather map connect places with the same temperature.
4. Air pressure increases at higher elevations.
5. A front that forms when a warm air mass meets a cold air mass and no movement occurs is an occluded front.
6. When cirrus clouds form close to the ground, the result is ground fog.
7. The Coriolis effect causes all winds in the Northern Hemisphere to curve to the left.
8. Water vapor, or moisture, enters the air through the process of evaporation.

Concept Mapping

Complete the following concept map for Section 1–1. Refer to pages K6–K7 to construct a concept map for the entire chapter.

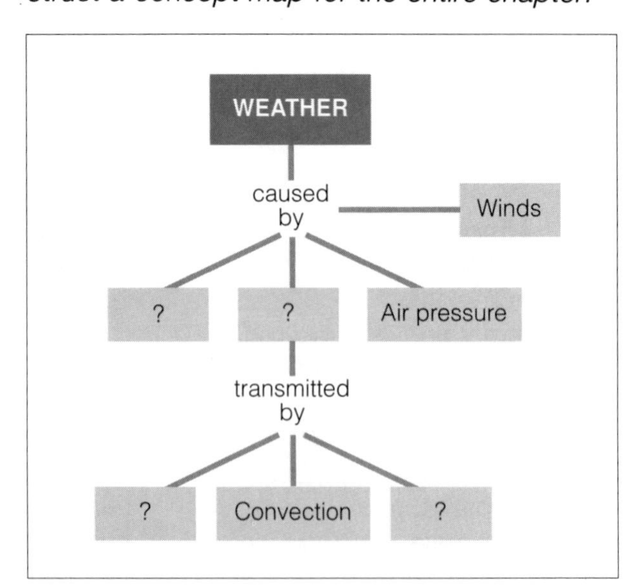

Concept Mastery

Discuss each of the following in a brief paragraph.

1. How are rain, sleet, snow, and hail formed?
2. How is weather information from a particular observation station shown on a weather map? What kinds of weather data are recorded on a weather map?
3. Describe the three main types of clouds.
4. What is the difference between a cyclone and an anticyclone?
5. Describe the greenhouse effect. What is its importance?
6. Compare how heat energy is transferred by conduction, by convection, and by radiation.
7. Describe the formation of a hurricane.
8. Explain how the Coriolis effect determines the direction of global winds.

Critical Thinking and Problem Solving

Use the skills you have developed in this chapter to answer each of the following.

1. **Making calculations** Suppose that 1 kilogram of air can hold 15 grams of water vapor but actually holds only 10 grams. What is the relative humidity? How much water vapor is in the air if the relative humidity is 100 percent? What if the relative humidity is 30 percent?
2. **Interpreting a map** Use the weather map in Figure 1–39 on page 46 to answer the following questions. Find the isobar passing through New York. What is the air pressure in New York? What other locations have the same air pressure as New York? What type of front is approaching Kansas City? Predict the probable weather conditions in Kansas City the day after this map was issued.
3. **Applying concepts** Another instrument that can be used to measure relative humidity is called a hair hygrometer. A hair hygrometer can be read directly without using a chart. It works on the principle that human hair changes length in proportion to changes in relative humidity. Hair gets longer as relative humidity increases. How is this related to the fact that people with curly hair find that their hair gets "frizzy" in humid weather?
4. **Making inferences** Is it possible for both the temperature and the amount of water vapor in the air to change while the relative humidity remains the same? Why or why not?
5. **Relating concepts** The diagram shows how the air pressure changes toward the center of a hurricane. Does the air pressure increase or decrease toward the center of the hurricane? How is your answer related to what you know about hurricanes and how they form?

   1000
   1004
   1008
6. **Making inferences** Explain why an aneroid barometer can be used to measure elevation as well as air pressure.
7. **Using the writing process** Pretend that you are the captain of a nineteenth-century sailing ship. You are preparing to make a voyage from the east coast of the United States to England and back again. Write a letter to a friend or relative describing the route you will take in order to make the best use of global wind patterns. (You may wish to refer to the diagram in Figure 1–16.)

What Is Climate?

Guide for Reading

After you read the following sections, you will be able to

2–1 What Causes Climate?

- Explain what is meant by climate.
- Describe how temperature and precipitation influence climate.

2–2 Climate Zones

- Identify the main characteristics of the Earth's major climate zones.

2–3 Changes in Climate

- Describe how the Earth's climate has changed.
- Explain how changes in climate have affected living things.

Just imagine that you can travel into the distant past or the far-off future by means of a time machine. You enter your time machine in the midwestern United States at the end of the twentieth century. What will you find as you step out into the world of 6000 years ago?

The air is warm. But it is also much more humid than you are used to. Tall grasses grow everywhere. Many different types of flowers add splashes of brilliant color. The world you have reached is a tropical grassland. Obviously, the climate 6000 years ago was markedly different from the climate today. What type of climate might you find farther back in time?

Returning to the time machine, you set the controls for 16,000 years ago. When you step out of the machine again, a blast of bitter-cold air hits you. Stretching away to the north is a vast sheet of ice. In the distance, a herd of woolly mammoths grazes in a bleak, snow-covered landscape. You quickly retreat to the warmth of the time machine! As you set the controls for 30,000 years in the future, you think about how drastically the climate has changed over thousands of years. What might await you in the future?

Journal *Activity*

You and Your World What does the word climate mean to you? Hot? Cold? Wet? Dry? In your journal, briefly describe the climate in the area where you live. Then describe how you think the climate in your area might be different in the future. How might it have been different in the past?

Woolly mammoths grazed the snow-covered grasslands of the midwestern United States about 16,000 years ago.

Guide for Reading

Focus on these questions as you read.

- *What are the factors that determine climate?*
- *What are the factors that influence temperature and precipitation?*

2–1 What Causes Climate?

On your way to school today, you may have made some observations about the weather. Even before you left your house, you were probably aware of two important weather factors: temperature and precipitation. You may have looked at an outdoor thermometer to check the morning temperature. Was the air cold, cool, warm, or hot? You also may have checked the precipitation. Was it raining or snowing?

If you were to keep a record of the weather in your area for an extended period of time, you would discover some general conditions of temperature and precipitation (rain, snow, sleet, hail). Such general conditions are described as the average weather for your area. **Climate** is the name for the general conditions of temperature and precipitation for an area over a long period of time. Every place on Earth has its own climate. For example, the climate of the southwestern United States tends to be hot and dry all year. The climate of Florida is also hot, but it is much wetter than the climate of the Southwest. What is the climate like where you live?

The climate of any region on the Earth is determined by two basic factors: temperature and precipitation. Different combinations of temperature and precipitation are used to classify the Earth's major climates. Temperature and precipitation are in turn influenced by several other factors.

Figure 2–1 *Climate is all the characteristics of the weather in an area over a long period of time. Some parts of the world are dominated by bleak, snow-capped mountain ranges. White sandy beaches and palm trees are the rule in tropical regions.*

Factors That Affect Temperature

Latitude, elevation, and the presence of ocean currents are three natural factors that affect the temperature at a particular location. The extent to which these factors influence climate varies from place to place.

LATITUDE Latitude is a measure of the distance north and south of the equator. (Recall from Chapter 1 that the equator is the imaginary line which separates the Earth into two halves, or hemispheres.) Latitude is measured in degrees (°). Areas close to the equator, or 0° latitude, receive the direct rays of the sun. These direct rays provide the most radiant energy. So areas near the equator have a warm climate.

Farther from the equator, the sun's rays are not as direct. As a result, areas farther from the equator receive less radiant energy. So climates are cooler in latitudes farther north and south of the equator. In general, the lowest average temperatures occur near the poles (90° north and south latitude), where the sun's rays are least direct.

ELEVATION Elevation, or altitude, is distance above sea level. As elevation increases, the air becomes less dense. This means that there are fewer gas molecules in the air and they are spread farther apart. Less-dense air cannot hold as much heat as denser air. So as elevation increases, the temperature decreases. The temperature at the top of a mountain is lower than the temperature at sea level.

OCEAN CURRENTS An ocean current is a "river" of water that flows in a definite path in the ocean. Some ocean currents are warm water currents. Other ocean currents are cold water currents. The major warm currents and cold currents are shown in Figure 2–3.

The surface temperature of water affects the temperature of the air above it. Warm water warms the air and cold water cools the air. So land areas near warm water currents have warm temperatures.

Figure 2–2 *Although Mt. Kilimanjaro in Kenya, Africa, is located near the equator, snow is visible on top of the mountain. What factors other than elevation affect climate?*

Activity Bank

What Are Density Currents?, p.123

Figure 2–3 *This map shows the general direction of flow of the world's major warm and cold ocean currents. Is the Gulf Stream a warm current or a cold current?*

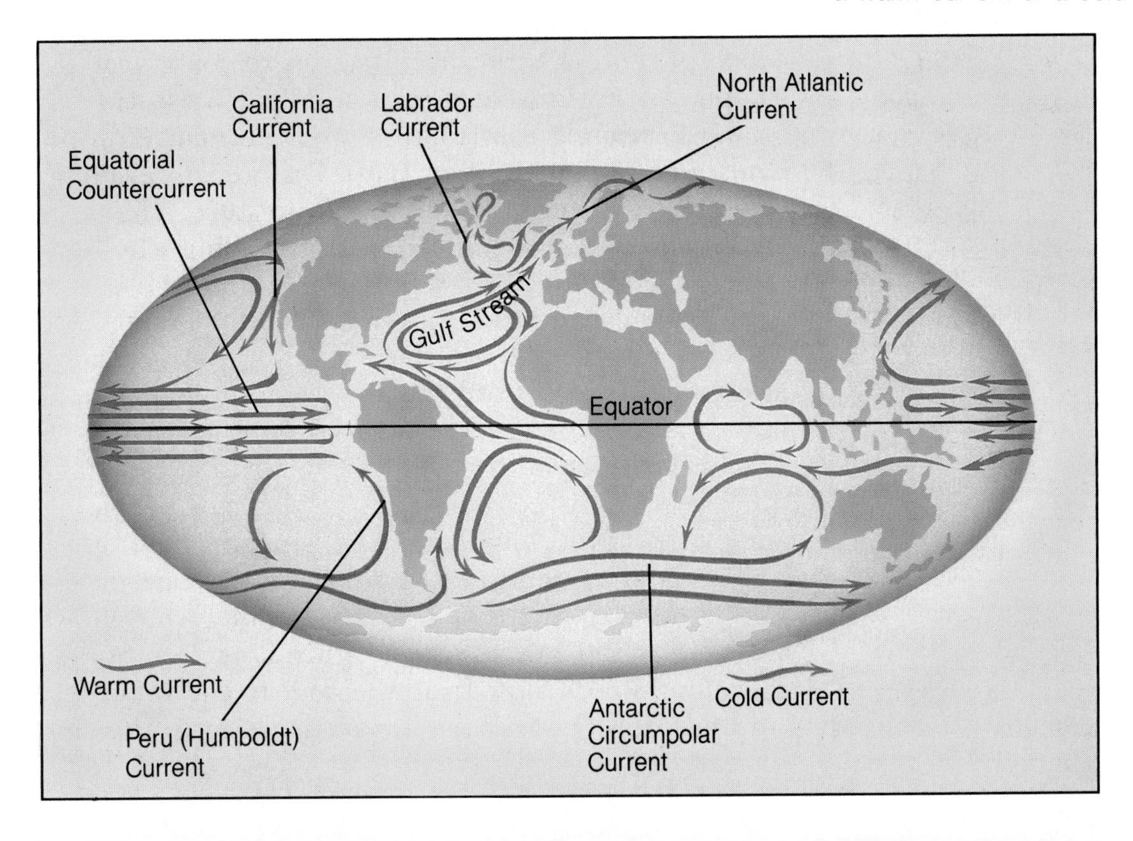

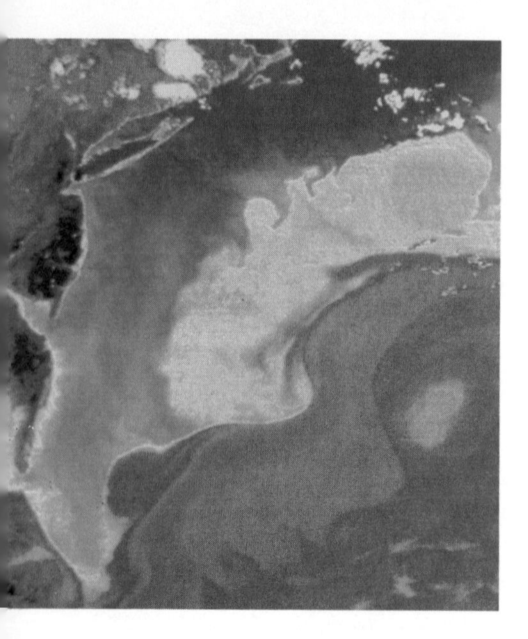

Figure 2–4 *The Gulf Stream appears purple in this satellite photograph. You can see the eastern coast of the United States on the left side of the photo.*

Land areas near cold water currents have cool temperatures.

Ocean currents traveling away from the equator are warm water currents. Land areas located near these currents have warm temperatures. The Gulf Stream is an ocean current that carries warm water from the southern tip of Florida along the eastern coast of the United States. How do you think the warm waters of the Gulf Stream affect the climate of the eastern United States?

Ocean currents traveling toward the equator are cold water currents. Areas located near these currents have cool temperatures. Off the western coast of the United States, the California Current flows toward the equator. What kind of temperatures would you expect along the coast in this area?

Factors That Affect Precipitation

The two natural factors that affect the amount of precipitation at a particular location are prevailing winds and mountain ranges. As with temperature factors, the effects of precipitation factors vary from place to place.

PREVAILING WINDS A wind that blows more often from one direction than from any other direction is called a **prevailing wind.** (Recall the global winds called prevailing westerlies that you read about in Chapter 1. From what direction do these winds blow?) Prevailing winds have a great influence on the climate of regions in their path. Different prevailing winds carry different amounts of moisture. The amount of moisture carried by a prevailing wind affects the amount of precipitation a region receives.

Warm air can hold more moisture than cold air. As warm air rises, it cools and cannot hold as much moisture. The moisture the air can no longer hold falls to the Earth as some form of precipitation. Thus winds formed by rising warm air tend to bring precipitation. As cold air sinks, it becomes warmer and can hold more moisture. So winds formed by sinking cold air tend to bring little precipitation.

The direction from which a prevailing wind blows also affects the amount of moisture it carries. Some prevailing winds blow from the land to the water (a land breeze). Others blow from the water to the land

ACTIVITY WRITING

Factors Affecting Climate

The climate of a region is determined by several different factors. Pretend that you live in one of the cities listed below. Write a letter to a pen pal in another country describing the climate in your city. Use library reference materials in writing your description. Be sure to explain which factors are more important than others and why.

Juneau, Alaska
Honolulu, Hawaii
Tokyo, Japan
Athens, Greece
Santiago, Chile

(a sea breeze). Which kind of prevailing wind do you think carries more moisture? Remember that moisture gets into the air as a result of the evaporation of water from the Earth's surface. Where is there more water? The prevailing winds that blow from the water carry more moisture than those that blow from the land. So areas in the path of a prevailing wind that originates over water usually receive a lot of precipitation. How much precipitation do you think areas receive in the path of a prevailing wind blowing from inland?

A region that receives a very small amount of precipitation (less than 25 centimeters of rainfall a year) is called a desert. The combined effect of a prevailing wind's moisture content and its direction can make it possible for a desert to exist near a large body of water.

Let's see how this happens. The Sahara in northern Africa is the largest desert on Earth. (In fact, the name sahara is the Arabic word for desert.) It is also one of the driest places on Earth. Yet the Sahara is bordered on the west by the Atlantic Ocean! However, the prevailing winds that blow across the Sahara (and carry little moisture) originate far inland, where they are caused by sinking cold air (which brings little precipitation). These two factors combine to make the prevailing winds over the Sahara very dry. As a result, little precipitation reaches the Sahara, even though a large body of water is nearby.

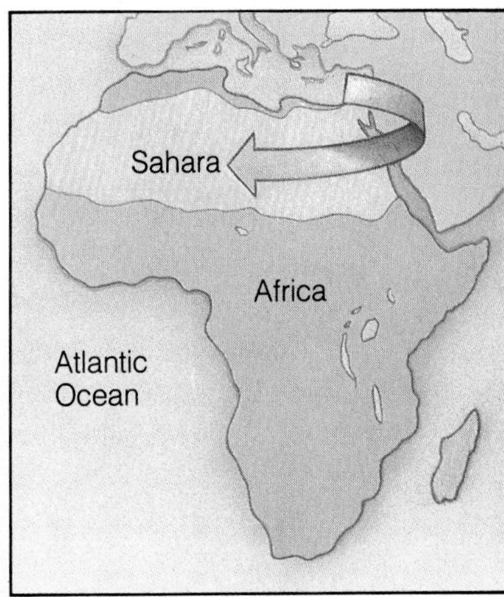

Figure 2–5 *Although the Sahara in Africa is bordered on the west by the Atlantic Ocean, it is one of the driest places on Earth! The prevailing winds, which originate far inland, carry little moisture as they sweep south and then west across the region.*

To get a better idea of how dry the Sahara really is, pour water into a graduated cylinder up to the 25-centimeter mark. This is the total amount of water that falls on the Sahara in one year. In some parts of the Sahara, no rain at all has fallen for more than 20 years!

MOUNTAIN RANGES The amount of precipitation at a particular location is also affected by mountain ranges. A mountain range acts as a barrier to prevailing winds. As you can see from Figure 2–6, mountains cause air to rise. As the air rises, it cools. Remember that cold air cannot hold as much moisture as warm air. So the moisture in the rising air falls to the Earth as precipitation. As a result, the **windward side** of a mountain, or the side facing toward the wind, receives a great deal of precipitation. The region on the windward side of a mountain has a wet climate.

Conditions are far different on the **leeward side** of a mountain, or the side facing away from the wind. By the time the prevailing winds reach the top of the mountain, they have lost most of their moisture in the form of precipitation. So relatively dry air moves down the leeward side of the mountain. As

Figure 2–6 *There is a rainy climate on the windward slopes of a mountain range because moist air rises, cools, and forms rain clouds. Dry air moving down the leeward slopes results in a desert climate.*

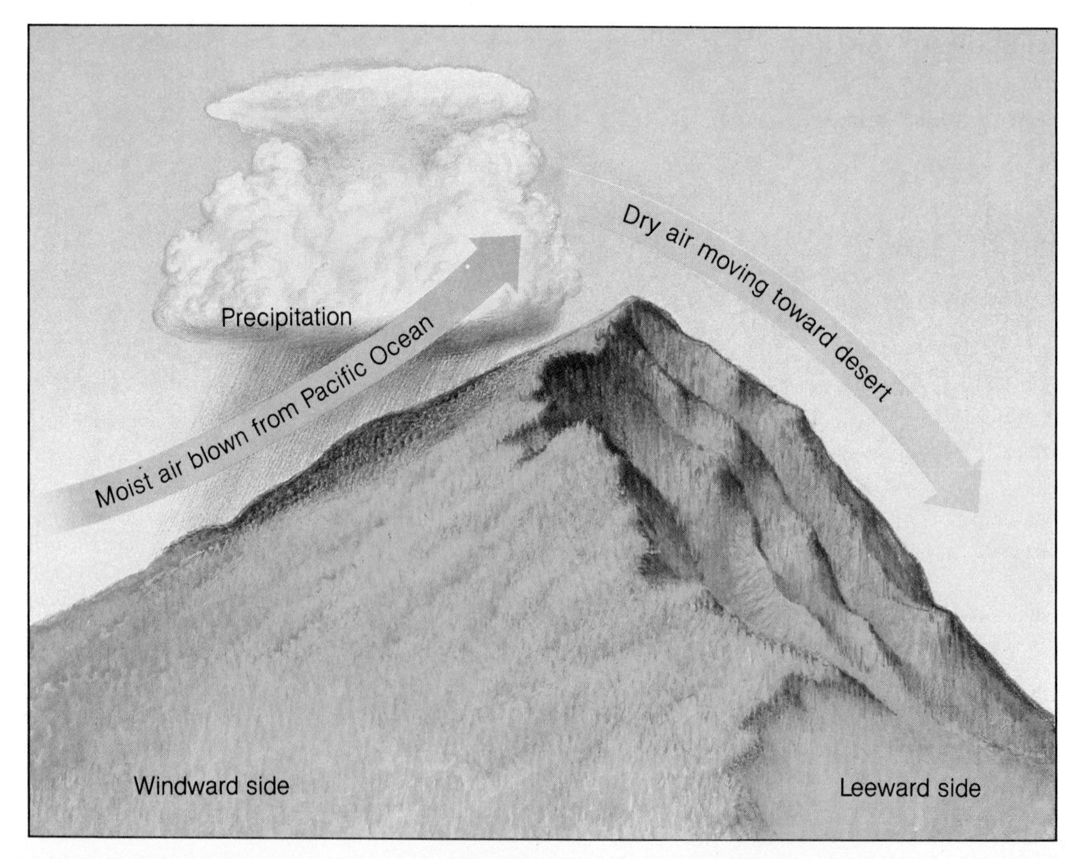

a result, there is little precipitation on the leeward side of a mountain. The area on the leeward side of a mountain is called a rain shadow. What kind of climate does this area have? You are correct if you said this area has a dry climate. In fact, there are usually dry areas called rain-shadow deserts on the leeward side of a mountain range.

On the west coast of the United States is a mountain range called the Sierra Nevadas. Areas to the west of the Sierra Nevadas (the windward side) receive a large amount of precipitation from the prevailing winds, which blow in from the Pacific Ocean. Land areas east of the Sierra Nevadas (the leeward side) receive little precipitation because the prevailing winds have lost most of their moisture by the time they cross the mountain range. The result is a rain-shadow desert called the Great Basin on the eastern side of the Sierra Nevada mountain range. The Great Basin extends south from Washington State into Nevada and Utah.

Figure 2–7 *The western side of the Sierra Nevada mountain range has a moist climate (top). The Great Basin in Utah, on the eastern side of the Sierra Nevadas, has a desert climate (bottom). Which is the windward side? The leeward side?*

2–1 Section Review

1. What two factors determine climate? What conditions influence these factors?
2. Describe how cold water currents and warm water currents affect the climate in locations near these currents.
3. Explain the following conditions:
   a. The peak of a mountain near the equator is covered with snow throughout the year.
   b. Deserts are located on the eastern side of the Rocky Mountains.

Critical Thinking—*Applying Concepts*

4. Suppose that you live in a coastal region on the windward side of a mountain range. A warm water current flows along the coast. Describe the climate in your region.

Guide for Reading

Focus on this question as you read.

▶ *How are the Earth's climates classified into major climate zones?*

2–2 Climate Zones

An Alaskan Eskimo trudges on snowshoes through the frozen wasteland above the Arctic Circle. Nearby, a polar bear hunts seals in the icy-cold Arctic Ocean. Thousands of kilometers farther south, tourists wander through the steamy rain forests of Hawaii. Exotic tropical birds call to each other from the dense treetops. Why do Alaska and Hawaii have such different climates?

The Earth's climates can be divided into general climate zones according to average temperatures. These climate zones can be broken down into subzones. Even the subzones have further subdivisions. In fact, scientists even classify very localized climates as **microclimates.** A microclimate can be as small as your own backyard!

The three major climate zones on the Earth are the polar, temperate, and tropical zones. Temperatures in these three climate zones are determined mainly by the location, or latitude, of the zone. Figure 2–8 shows the locations of the three major climate zones. In what climate zone is Alaska located? Hawaii? In what climate zone do you live?

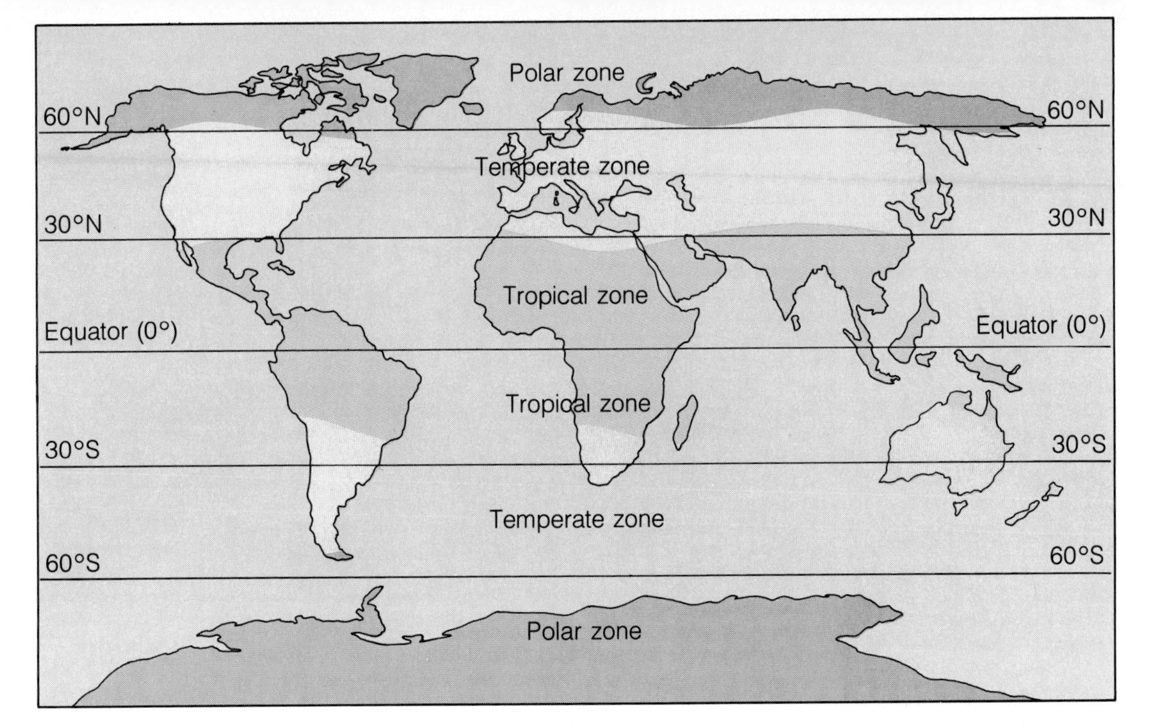

Figure 2–8 *The Earth's three major climate zones are shown in this diagram. In which zone is most of the United States located?*

Polar Zones

In each hemisphere, the **polar zone** extends from the pole (90°) to about 60° latitude. Polar climates have the coldest average temperatures. Within the polar zones, the average yearly temperature remains below freezing (below 0°C). Polar climates have no summer. Even during the warmest months of the year, the average temperature does not rise above 10°C. There is little precipitation in the polar zones.

Polar zones are also known as high-latitude or arctic climates. The polar zones include the icecaps of Greenland in the Northern Hemisphere and Antarctica in the Southern Hemisphere. These icecaps remain frozen throughout the year. However, there are some places in the polar zones where the snow melts during the warmest part of the year. The northern coasts of Canada and Alaska and the southern tip of South America are examples of these places.

Activity Bank

What Is Your Latitude?, p.124

Figure 2–9 *Adelie penguins enjoy a brisk swim in the cold waters off Antarctica, which is in the polar climate zone.*

Temperate Zones

In each hemisphere, the **temperate zone** is found between 60° and 30° latitude. In the areas of the temperate zones farther from the equator, snow is common in the winter. In the areas of the temperate

Figure 2–10 *The yellow-bellied racer is one of several snakes that make their home in Arizona's Sonoran Desert.*

zones closer to the equator, rain normally falls all year round. But the average amount of precipitation is about the same throughout the temperate zones. Average temperatures in the temperate zones vary greatly. They range from about 5°C to more than 20°C. These temperatures fall between those of the polar and the tropical zones.

Temperate zones, or middle-latitude climates, cover a huge portion of the Earth. So the temperate zones can include the cool rain forests of Washington State as well as the hot rain forests of southern China, with many different climates in between. Most of the United States is in the temperate zone.

Deserts in the temperate zones are usually located inland, far away from the oceans. The winds that blow across these inland deserts carry little moisture. Inland deserts are found in Australia (the Great Sandy Desert) and Central Asia (the Gobi Desert).

Many people mistakenly believe that temperate deserts are always hot. Certainly this is true of deserts during the day. But at night, the temperature in the desert can drop to below freezing! How is this possible? The low humidity and cloudless skies allow a tremendous amount of radiant energy to reach the ground and heat it during the day. But these same conditions also allow the heat to escape rapidly at night, causing the temperature to drop dramatically. As a result, temperatures in the desert can range from 20°C at 2 o'clock in the afternoon to 0°C at 2 o'clock in the morning.

Examining a Microclimate

1. Locate two separate areas in your neighborhood that appear to have different microclimates.

2. Carefully observe each area for several days.

■ Describe the similarities and differences in your two microclimates. What factors cause these differences? How is a microclimate related to the Earth's major climate zones?

Another common misunderstanding people have about deserts is that they are barren and lifeless. However, several kinds of plants and animals are able to live in the desert. For example, in the Sonoran Desert of the southwestern United States and Mexico, plants such as sagebrush and giant saguaro cacti grow. Animals such as lizards, snakes, and cougars also live in this desert.

Radiant Energy and Climate

Radiant energy from the sun strikes different areas of the Earth at different angles. Using a flashlight and a piece of paper taped to the wall, design an experiment to show how the angle of sunlight affects the intensity of the light.

■ Relate your observations to the three major climate zones of the Earth.

Tropical Zones

The **tropical zones,** which extend from 30° north and south latitude to the equator (0°), have high temperatures and high humidity. Precipitation in the tropical zones is usually very heavy during part of the year. Tropical zones are also known as low-latitude climates.

Tropical climates have the warmest average yearly temperatures. There is no winter in tropical climates. In a tropical climate, the average temperature during the coldest month of the year does not fall below 18°C.

In the tropical zones, many deserts are located on the western coasts of continents. This is because the prevailing winds in the tropical zones (the northeast and southeast trades) blow from east to west. High mountains along the western coast of a continent block these prevailing winds from reaching the coast. Rain falls on the eastern (windward) side of

Figure 2–11 *When you think of the tropical climate zone, images such as the lush Hawaiian rain forest probably come to mind. But the barren Atacama Desert, a cold desert in Peru, is also in the tropical zone.*

CAREERS

Paleoclimatologist

The study of fossils is one way to gather information about climates of the past. Scientists who study fossils and other natural sources to learn about ancient climates are **paleoclimatologists.** (The prefix *paleo-* means ancient.) Their work has shown that the Earth's history has included a series of major glaciations, or cold periods, falling between interglacials, or warm periods.

If you enjoy solving scientific puzzles and working outdoors, you might be interested in a career in paleoclimatology. For more information, write to the American Geological Institute, 4220 King Street, Alexandria, VA 22302.

the mountains. Areas on the western (leeward) side of the mountains do not receive much rainfall and thus become deserts. These deserts are often cold deserts due to the presence of cold ocean currents along the western coasts of the continents. For example, the Atacama Desert in parts of Chile and Peru is a cold desert located on the western coast of South America.

Marine and Continental Climates

Within each of the three major climate zones there are **marine climates** and **continental climates.** Areas near an ocean or other large body of water have a marine climate. Areas located within a large landmass have a continental climate.

Areas with a marine climate receive more precipitation than areas with a continental climate. Can you explain why? Temperatures in areas with a marine climate do not vary greatly. Areas with a marine climate have warm (not hot) summers and mild winters. This is because their nearness to a large body of water has a moderating effect on the air temperature.

A continental climate is drier than a marine climate. Why? There is usually a great range in average temperatures during the year. Areas with a continental climate have hot summers and cold winters. Most of the world's deserts that are located just north and south of the equator have a continental climate.

The Four Seasons

As you have just read, the Earth has three major climate zones: polar, temperate, and tropical. Most places on the Earth also have four distinct seasons: winter, spring, summer, and autumn. The different seasons are caused by the tilt of the Earth's axis. The Earth's axis is an imaginary line through the center of the Earth. The Earth turns, or rotates, on this axis from west to east once every 24 hours.

The Earth's axis is not straight up and down. Instead, it is tilted at an angle of 23.5°. So as the Earth moves around the sun, or revolves, the axis is tilted away from the sun for part of the year and toward the sun for part of the year. When the

Northern Hemisphere is tilted toward the sun, that half of the Earth has summer. At the same time, the Southern Hemisphere is tilted away from the sun and has winter. So on a particular day, it may be summer in San Francisco, California, but winter in Sydney, Australia. Which hemisphere of the Earth is tilted away from the sun when the Southern Hemisphere has summer? Which hemisphere is tilted toward the sun?

How does the tilt of the Earth's axis cause summer and winter? Look at Figure 2–12. The hemisphere that is tilted toward the sun receives more direct rays than the hemisphere that is tilted away from the sun. Recall from Chapter 1 that the greatest heating occurs when the sun's rays are most direct. Think about a typical sunny day. The sun's rays feel hotter at noon, when the sun is almost directly overhead, than they do in the late afternoon, when the sun is low in the sky. So the Earth's land surface, oceans, and atmosphere receive more heat in the hemisphere that is tilted toward the sun. The

A Seasonal Journal

Read *Circle of the Seasons* by the American naturalist Edwin Way Teale (1899–1980).

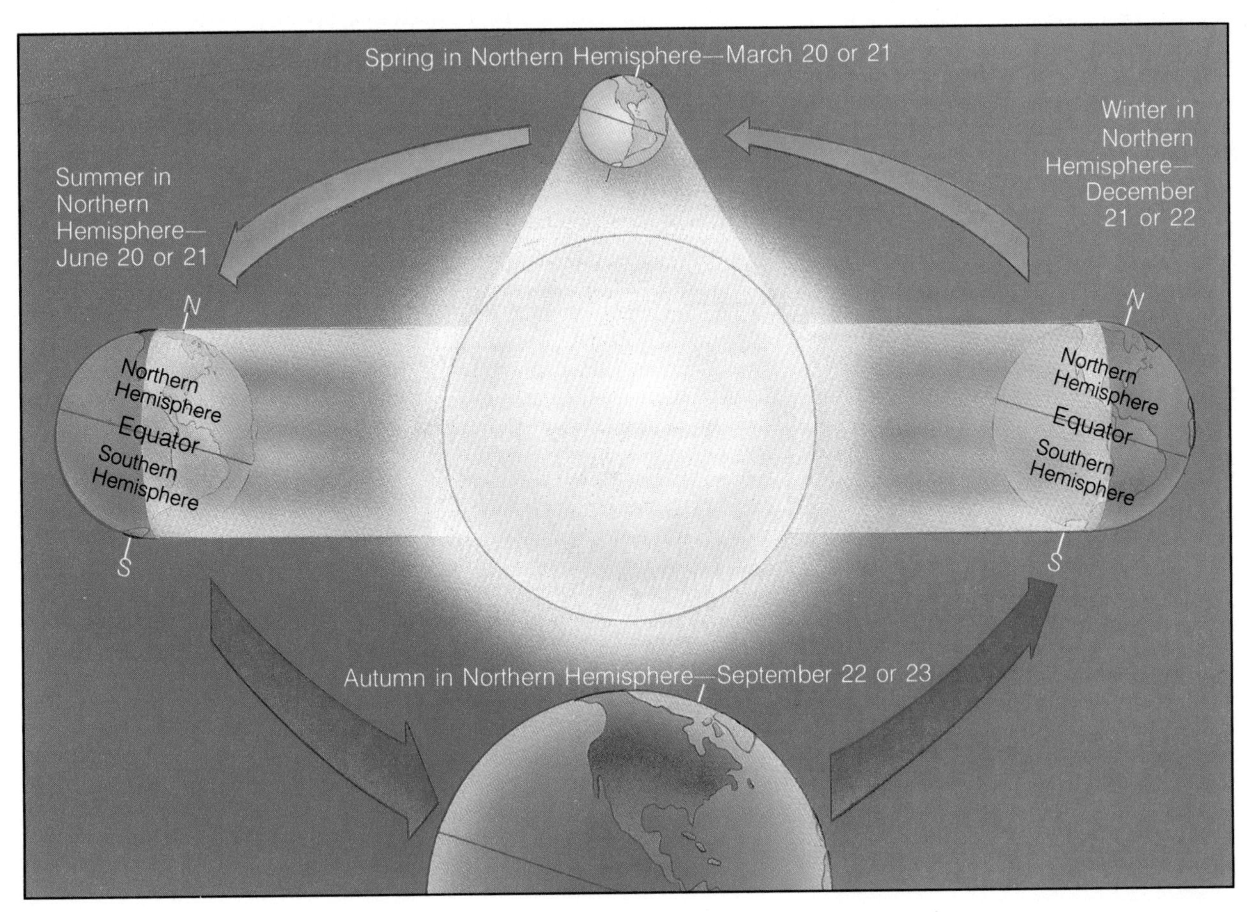

Figure 2–12 *When the North Pole is tilted toward the sun, the Northern Hemisphere receives the direct rays of the sun. It is summer. When the North Pole is tilted away from the sun, the Southern Hemisphere receives the direct rays of the sun. What season is it in the Northern Hemisphere?*

Figure 2–13 *These photographs were taken on the same day in the Northern and Southern hemispheres. It is winter in Chicago, Illinois. But it is summer on the golf course in Australia. On this day, is the North Pole tilted toward or away from the sun?*

result is the summer season. Just the opposite happens in the hemisphere that is tilted away from the sun. This hemisphere receives slanting rays from the sun and therefore less heat. The result is the winter season.

Twice during the year neither hemisphere is tilted toward the sun. This occurs in spring and autumn. In the Northern Hemisphere, spring begins on March 20 or 21 and autumn begins on September 22 or 23. What season begins in the Southern Hemisphere when spring begins in the Northern Hemisphere?

ACTIVITY DISCOVERING

Temperature, Daylight, and the Seasons

1. Keep a record of the high and low temperatures for each day of the school year.

2. Record the time at which the sun rises and sets each day as given in the daily newspaper.

3. Calculate the number of hours of daylight for each day.

■ What happens to the high and low temperatures and the length of daylight from season to season during the year?

■ Is there a relationship between the temperature and the amount of daylight? Explain.

2–2 Section Review

1. What are the Earth's three major climate zones?
2. Describe the location of each major climate zone. What conditions of temperature and precipitation are typical of each zone?
3. What is the difference between a marine climate and a continental climate?
4. Why is it summer in the Southern Hemisphere when it is winter in the Northern Hemisphere?

Critical Thinking—*Relating Concepts*

5. Explain in terms of radiant energy why polar zones have the coldest average temperature and tropical zones have the warmest average temperature.

2–3 Changes in Climate

Guide for Reading

Focus on this question as you read.

- *What causes the changes in the Earth's climate?*

You know from experience that weather changes from day to day. Sometimes weather seems to change from hour to hour! Climate, however, seems to remain relatively unchanged. But climates do change slowly over time. In fact, the climate of a region can change from a temperate rain forest to a tropical desert within a relatively short time in the Earth's history. (Remember, however, that the Earth is about 4.6 billion years old. So a "short time" in the Earth's history could be thousands or even millions of years!)

What causes climates to change? Major climate changes may be caused by one or more of three natural factors. **The three natural factors responsible for climate changes are the slow drifting of the continents, changes in the sun's energy output, and variations in the position of the Earth relative to the sun.** These "natural" factors are not related to human activities. However, the results of human activities, which include increased atmospheric levels of carbon dioxide caused by burning fossil fuels, may also lead to changes in climate.

As you might guess, major climate changes have a tremendous impact on the Earth and on the organisms that inhabit it. Just think of the changes in climate you saw during your imaginary trip into the

Figure 2–14 *Climate changes may have caused the extinction of the dinosaurs 65 million years ago. But were those changes brought about by the slow drifting apart of the continents or by a sudden, catastrophic meteor impact?*

past. Major climate changes that occurred in the past have had dramatic effects on the Earth, including a series of ice ages and perhaps the extinction (dying out) of the dinosaurs. Scientists have not yet determined the causes of past climate changes. Once they do, they will be better able to predict future climate changes and their effects on the Earth and its living things, including humans.

Ice Ages

From time to time throughout the Earth's history, much of the Earth's surface has been covered with enormous sheets of ice. Such periods are called ice ages. Scientists have found evidence of at least four major ice ages during the last 2 million years. Earth scientists call these ice ages **major glaciations.**

During an ice age, or major glaciation, the average temperature on the Earth was about 6°C below the average temperature today. Each glaciation lasted about 100,000 years or more. The most recent glaciation began about 1.75 million years ago and ended only about 10,000 years ago. (Remember, 10,000 years is like the blink of an eye compared with the age of the Earth!) During the last glaciation,

Figure 2–15 *Geological evidence indicates that the white areas on this map were covered with huge masses of ice during the last ice age. What is another name for an ice age?*

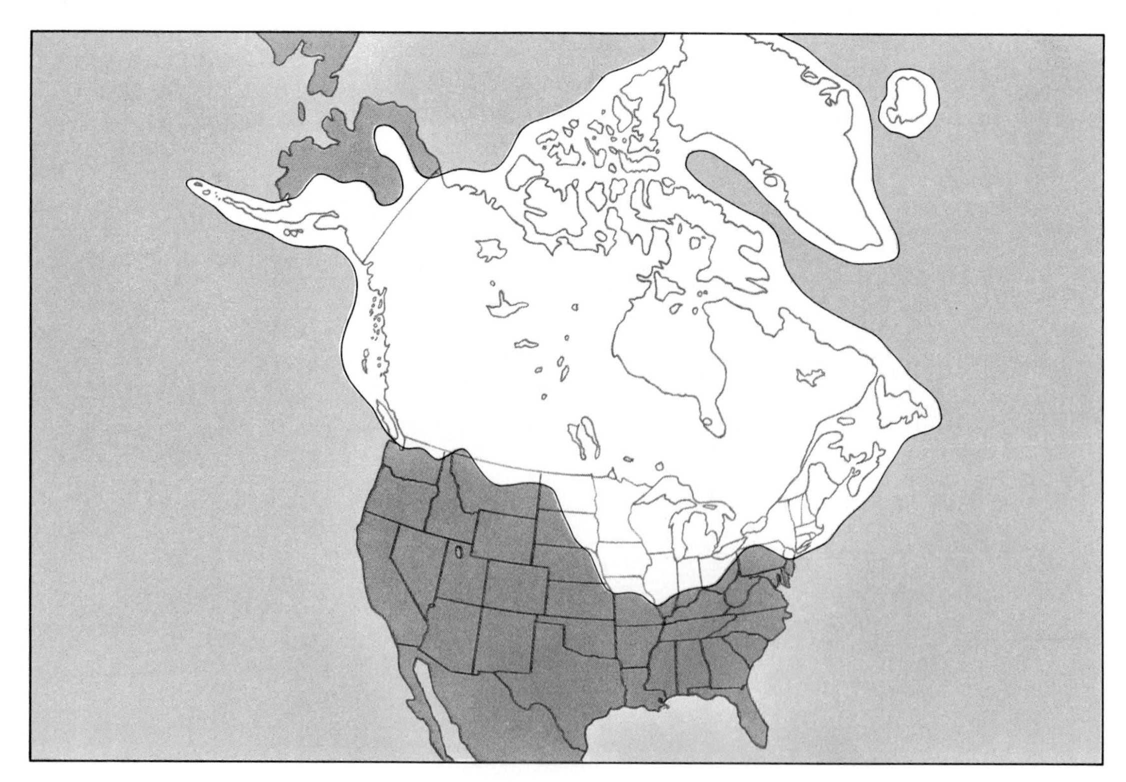

a great sheet of ice covered the United States as far south as Iowa and Nebraska. So much water was locked in the ice that the average sea level rose about 85 meters when the ice melted. That is enough water to cover a 20-story building!

The time periods between major glaciations are called **interglacials.** Interglacials are warm periods. During an interglacial, the average temperature was about 4° to 6°C higher than the average temperature during a major glaciation. A cold period, often called the Little Ice Age, lasted from 1500 to 1900. The Earth is now in an interglacial.

Although there are many theories about the cause of ice ages, the exact causes are not known. However, major glaciations are probably associated with gradual changes in the tilt of the Earth's axis and in variations in the shape of the Earth's path, or orbit, around the sun. Based on what you know about the sun's radiant energy, how could these changes influence the Earth's climate?

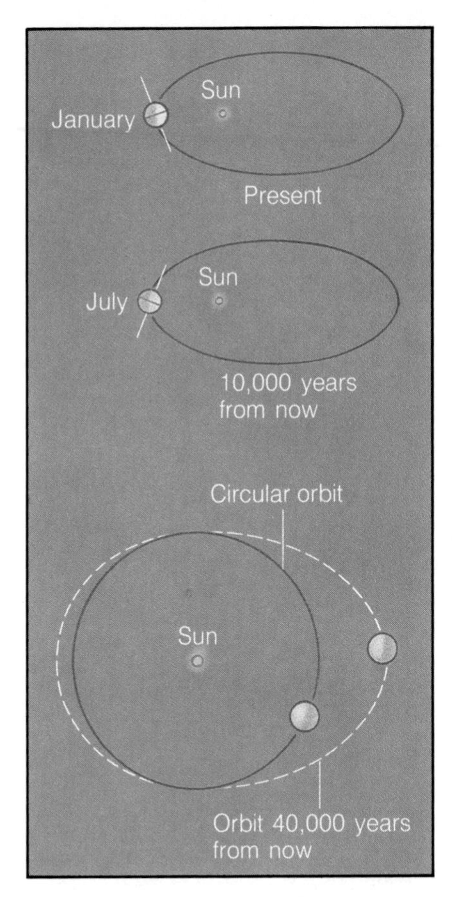

Figure 2–16 *The Earth's tilted axis changes direction like a spinning top. As a result, the time at which the Earth is closest to the sun gradually changes over a period of 10,000 years (top). The shape of the Earth's orbit is also changing—from nearly circular to slightly elliptical and back again (bottom). How might these changes affect the Earth's climate?*

Activity Bank

Earth's Elliptical Orbit, p.125

Drifting Continents

About 230 million years ago, all the Earth's landmasses were joined in one supercontinent. About 160 million years later, this supercontinent had broken apart and the individual continents had drifted close to their present locations.

The slow drifting apart of the continents caused dramatic climate changes. As the continents moved toward their present-day locations, the sea level dropped, volcanoes erupted, and much of the

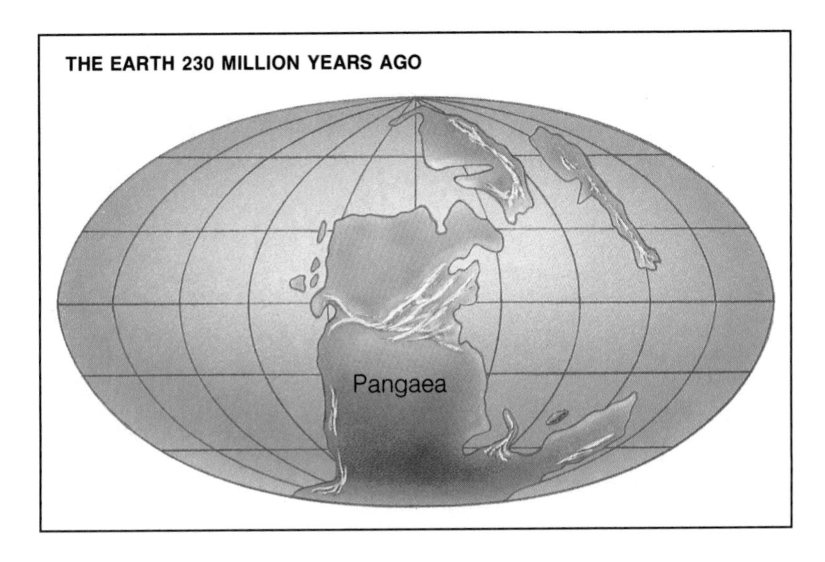

Figure 2–17 *Over millions of years, the shapes and positions of the Earth's continents have changed. What was the name of the supercontinent that existed 230 million years ago?*

Earth's surface was pushed upward. The combined effect of all these changes was a drop in temperature and precipitation all over the Earth. Because the continents move only a few centimeters per year, the climate changes caused by continental drift are very gradual and happen only over millions of years.

Extinction of the Dinosaurs

The climate changes caused by the drifting apart of the continents may have resulted in the extinction, or dying out, of the dinosaurs. About 65 million years ago, dinosaurs—and most other kinds of animals and plants—became extinct. Scientists do not know exactly what caused this mass extinction. Some biologists think it was caused by the slow process of climate change as a result of continental drift. Dinosaurs could not adapt fast enough to these drastic climate changes and died out. Also, many types of plants became extinct as a result of the climate changes. Without plants to eat, the plant-eating dinosaurs died out. And without the plant-eating dinosaurs as a food source, the meat-eating dinosaurs died out as well.

The theory of gradual climate change caused by drifting continents is only one explanation for dinosaur extinction. Many scientists think that the climate changes which caused the mass extinction

Figure 2–18 *Evidence found in different parts of the world indicates that a gigantic meteor collided with the Earth 65 million years ago. Many scientists think that the extinction of the dinosaurs was the direct result of this catastrophe.*

happened suddenly, rather than slowly. In 1978, the geologist Walter Alvarez and his father, Nobel prize-winning physicist Luis Alvarez, found evidence suggesting that a giant meteor or comet struck the Earth 65 million years ago. This collision resulted in a huge explosion. The explosion raised enormous clouds of dust and set off planetwide forest fires. As dust and smoke rose into the atmosphere, they may have blocked the sun's rays and caused the Earth's temperature to drop. The dinosaurs could not survive in the suddenly colder climate and died out.

Figure 2–19 *Carbon dioxide is released into the atmosphere by burning gasoline in automobile engines. What effect might a buildup of carbon dioxide have on the Earth's climate?*

Variations in Radiant Energy

Some scientists have tried to relate changes in the Earth's climate to changes in the sun's energy output. If the sun's energy output changes over time, these changes could have an effect on the Earth's temperature. During periods of high energy output, the Earth's temperature would rise. When the sun's energy output dropped, the Earth's temperature would fall. Although this seems logical, a relationship between variations in the sun's energy and climate changes on the Earth has not yet been demonstrated. In fact, scientists have found no evidence for any variations in the sun's energy output.

Global Warming

Humans have probably been altering the Earth's climate in some way ever since the discovery of fire. Only recently, however, have humans had a measurable effect on climate. In the mid-nineteenth century, industrialization led to the increased burning of fossil fuels. Fossil fuels include coal, oil, and natural gas. When these fuels are burned, they release carbon dioxide (CO_2) into the atmosphere. Recall from Chapter 1 that a buildup of CO_2 in the atmosphere results in a greenhouse effect. Like the glass in a greenhouse, CO_2 absorbs heat reflected from the Earth's surface and prevents the heat from escaping into space. As a result, the atmosphere becomes warmer.

Over the past 25 years, the amount of CO_2 in the atmosphere has increased by about 8 percent. By the middle of the next century, the percentage of CO_2 in

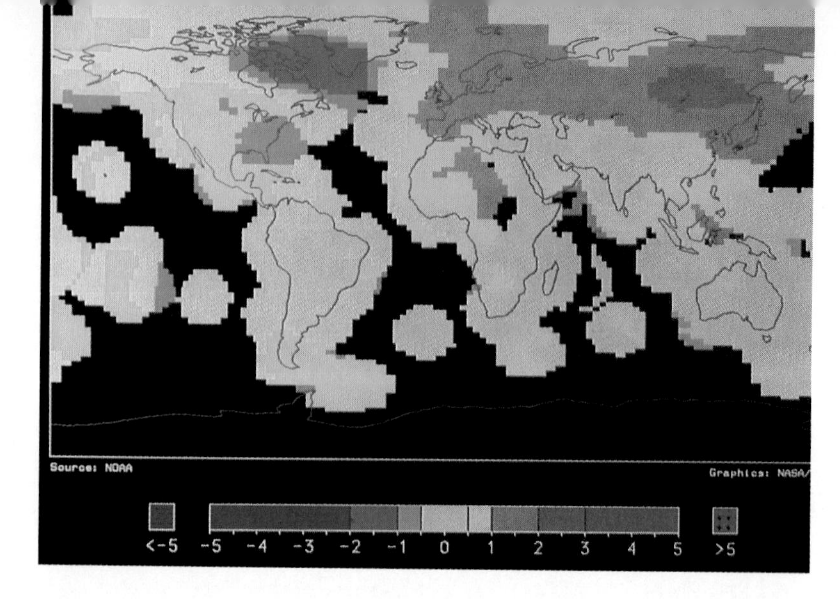

Figure 2–20 *This computerized chart shows average global temperatures for 1990. Blue represents the coldest temperatures and red represents the hottest.*

the atmosphere could be twice as much as it is today. How will this affect the Earth's climate? Climatologists have developed computer models to predict what will happen to the Earth's climate as a result of increased CO_2 levels. Most of these models predict an increase in temperatures of 1.5° to 4.5°C. These higher temperatures could lead to significant changes in the Earth's climate. Can you suggest what some of these changes might be?

Are we already beginning to feel the effects of global warming? Measurements made at weather stations around the world showed that the average surface temperature during 1990 was the highest in more than 100 years. However, scientists are not sure whether this warming trend resulted from an increase in CO_2 in the atmosphere. Yet many scientists and environmentalists recommend that people reduce their use of fossil fuels and thus limit the amount of CO_2 that escapes into the atmosphere. How can you and your classmates help in this effort?

Figure 2–21 *The arrival of El Niño is unpredictable. But it usually results in surprising weather for some parts of the world. Unexpected weather may include floods in Louisiana and severe droughts in Africa.*

El Niño

Some short-term climate changes may be the result of changes in ocean currents and global winds. Ocean currents help transfer heat to the atmosphere. This process generates global winds. The global winds, in turn, help move the ocean currents. Any major change in an ocean current can cause a change in climate. El Niño is an example of such a change. Let's see why.

A cold current that flows from west to east across the southern part of the Pacific Ocean turns toward

Problem Solving

Analyzing a Feedback Mechanism

At an international conference on global warming, British scientists presented the following theory: If the temperature of the Earth rises significantly as a result of global warming, the speed of global winds will increase. Increased wind speed will cause a greater disturbance of ocean water, which in turn will cause larger clouds to form. The clouds might either increase or decrease global warming. This theory is an example of a feedback mechanism. In a feedback mechanism, one event causes a series of other events, which in turn influence the first event.

Imagine that you are a scientist at the conference. You have been given a copy of the diagram shown here, which illustrates the feedback mechanism described by the British scientists. It is your job to interpret the diagram and to analyze how this particular feedback mechanism might influence global warming.

Relating Cause and Effect

1. According to the diagram, what changes will occur in the atmosphere as a result of global warming? What do you think will cause these changes to occur?

2. How will the changes in the atmosphere affect the ocean? Explain.

3. How will the changes in the ocean affect cloud formation?

4. Based on your knowledge of weather and climate, how might larger clouds make the Earth warmer? How might larger clouds make the Earth cooler?

5. Do you think that this feedback mechanism will increase or decrease the effects of global warming? Why?

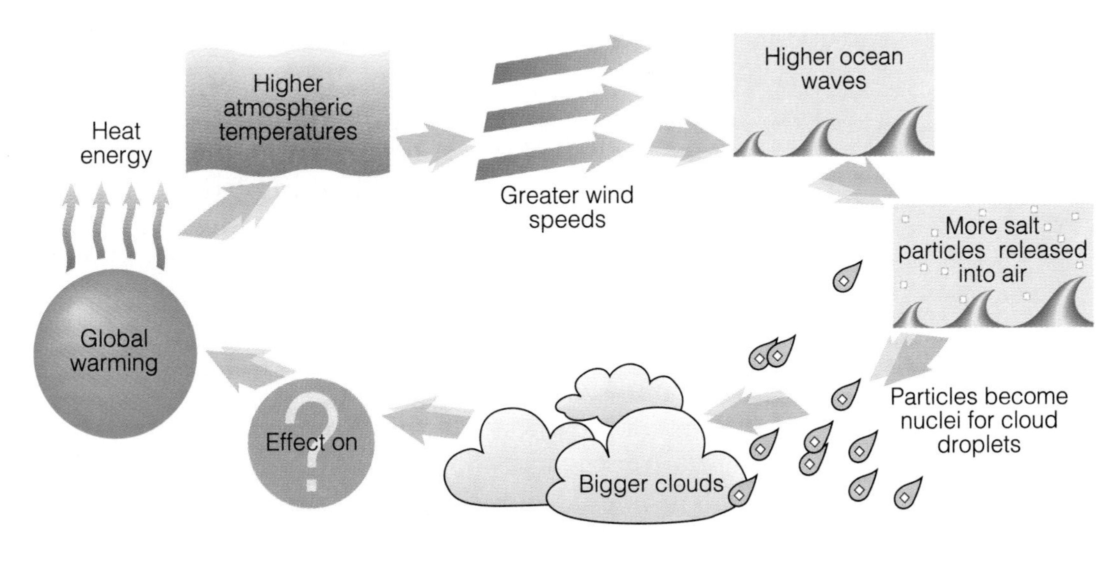

El Niño

Use reference materials in the library to find out more about El Niño. Write a brief essay, including answers to the following questions:

- What do scientists think may be the causes of El Niño?
- What are some of the worldwide weather changes that result from El Niño?
- Do scientists have any methods for predicting El Niño?
- What is upwelling? What effect does it have on ocean life in areas where it occurs?
- What effect does El Niño have on upwelling?

the equator along the coast of South America. As the current flows north along the coast of Chile and Peru, it is known as the Peru Current. Occasionally, the cold water of the Peru Current is covered by a thin "sheet" of warm water from the equator. Usually the warm water disappears fairly quickly. But every 2 to 10 years or so, strong winds spread the warm water over a large area. This unusual behavior of the Peru Current is known as El Niño.

El Niño, then, is a temporary current that arrives with little warning, usually around Christmas. (*El Niño* means the child in Spanish.) The warming caused by El Niño in the tropical zone results in dramatic changes in world climates. In 1982 and 1983, the strongest El Niño in history caused severe droughts in some regions. Other regions were subjected to unusually heavy rains and flooding. The extreme changes in climate resulted in more than 1000 deaths and much economic damage throughout the world.

Scientists have not yet discovered just what causes El Niño to appear. However, important progress has been made in understanding the interaction of the ocean and the atmosphere. Accurate predictions of future El Niños may be possible within a few years.

2–3 Section Review

1. What natural factors do scientists think may cause climate changes?
2. How might the slow drifting of the continents have caused the extinction of the dinosaurs? How might a collision with a meteor or comet have caused the extinction of the dinosaurs?
3. What is another name for an ice age? What is the period between ice ages called?
4. What is El Niño? What effect does it have on climate?

Connection—*Geology*

5. Some scientists think that volcanic eruptions may cause changes in the Earth's climate. How might volcanic dust in the atmosphere cause global temperatures to change? Would the resulting temperatures be lower or higher?

CONNECTIONS

The Birth of Agriculture

According to *archaeologists* Dr. Frank Hole and Joy McCorriston from Yale University, wild cereals were first domesticated about 10,000 years ago. (The word domesticate means to tame plants or animals to be able to use them for the benefit of humans.) This revolutionary event in human history took place in a region near the Dead Sea called the southern Levant. This region includes parts of the modern countries of Israel, Lebanon, and Jordan. The people of this region—who were called Natufians—were forced to learn how to grow cereals for food as a result of a change in climate.

About 12,000 years ago, the climate in the region became much hotter and dryer. As freshwater lakes dried up, the Natufians crowded into smaller and smaller areas where water was still available. The growing population led to food shortages. To reduce the food shortages, the Natufians began saving seeds from wild wheat, planting the seeds, and harvesting them for food.

The Natufians had previously relied only on the wild wheat for their food supply. But the wild wheat plants tended to scatter their seeds when they were cut, making them difficult to harvest. Then a genetic mutation, or change, in the wheat resulted in a few wheat plants with tougher stems that did not shed their seeds.

The Natufians found that the mutant wheat plants were easier to harvest than the wild plants. Gradually, more and more of the mutant wheat was selected and harvested. After about 20 years, only mutant plants were growing in the Natufians' wheat fields. The Natufians had domesticated wheat and thereby invented agriculture. So when you dig into a bowl of delicious, nutritious cereal, remember to thank the Natufians, who made it all possible!

Laboratory Investigation

Graphing Climate Information

Problem

How can you use temperature and precipitation data to classify the climates of cities in different parts of the world?

Materials *(per student)*

2 sheets of graph paper
3 different-colored pencils

Procedure

1. On a sheet of graph paper, plot the data for average monthly precipitation in Winnipeg, Canada.
2. On another sheet of graph paper, plot the data for average monthly temperature for Winnipeg.
3. Repeat steps 1 through 4 using the data for Izmir, Turkey, and for Ulan Bator, Mongolia.

Observations

1. Which city has the highest winter temperatures? The lowest?
2. Which city has the greatest temperature range from winter to summer? Which city has the smallest range?
3. Which city has the driest summers? The wettest?
4. Which city has the driest winters? The wettest?

Analysis and Conclusions

1. In which climate zone is each city located?
2. Which of the three cities has a marine climate? Which has a continental climate?
3. **On Your Own** How could you gather similar climate information about familiar locations? Choose three nearby cities and repeat this laboratory investigation using the climate information you gathered for those cities.

WINNIPEG, CANADA													
Month	**J**	**F**	**M**	**A**	**M**	**J**	**J**	**A**	**S**	**O**	**N**	**D**	**Year**
Temperature (°C)	–18	–16	–8	3	11	17	20	19	13	6	–5	–13	3
Precipitation (cm)	2.6	2.1	2.7	3	5	8.1	6.9	7	5.5	3.7	2.9	2.2	51.7
IZMIR, TURKEY													
Month	**J**	**F**	**M**	**A**	**M**	**J**	**J**	**A**	**S**	**O**	**N**	**D**	**Year**
Temperature (°C)	9	9	11	15	20	25	28	27	23	19	14	10	18
Precipitation (cm)	14.1	10	7.2	4.3	3.9	0.8	0.3	0.3	1.1	4.1	9.3	14.1	69.5
ULAN BATOR, MONGOLIA													
Month	**J**	**F**	**M**	**A**	**M**	**J**	**J**	**A**	**S**	**O**	**N**	**D**	**Year**
Temperature (°C)	–26	–21	–13	–1	6	14	16	14	9	–1	–13	–22	–3
Precipitation (cm)	0.1	0.2	0.3	0.5	1	2.8	7.6	5.1	2.3	0.7	0.4	0.3	21.3

Study Guide

Summarizing Key Concepts

2–1 What Causes Climate?

▲ The basic factors that determine climate are temperature and precipitation.

▲ Factors that affect temperature are latitude, elevation, and the presence of ocean currents.

▲ Factors that affect precipitation are prevailing winds and the presence of mountain ranges.

2–2 Climate Zones

▲ The Earth's three major climate zones are the polar, temperate, and tropical zones.

▲ Marine climates and continental climates occur within each of the three major climate zones.

▲ The four seasons are caused by the tilt of the Earth's axis.

2–3 Changes in Climate

▲ Natural factors that may cause changes in climate are continental drift, changes in the sun's energy output, and variations in the tilt of the Earth's axis and the shape of the Earth's orbit.

▲ The effects of human activities may lead to global warming.

Reviewing Key Terms

Define each term in a complete sentence.

2–1 What Causes Climate?

climate
prevailing wind
windward side
leeward side

2–2 Climate Zones

microclimate
polar zone
temperate zone
tropical zone
marine climate
continental climate

2–3 Changes in Climate

major glaciation
interglacial

Chapter Review

Content Review

Multiple Choice

Choose the letter of the answer that best completes each statement.

1. The climate in any region of the Earth is determined by temperature and
a. latitude. c. precipitation.
b. humidity. d. elevation.

2. A low-latitude climate is a climate that is
a. polar. c. arctic.
b. temperate. d. tropical.

3. The measure of distance north and south of the equator is called
a. altitude. c. latitude.
b. elevation. d. climate.

4. The climate zone with the coldest average temperatures is
a. tropical. c. temperate.
b. polar. d. marine.

5. The climate zone with the greatest temperature range is
a. tropical. c. temperate.
b. polar. d. marine.

6. Three factors that affect the temperature in an area are latitude, elevation, and
a. prevailing winds.
b. ocean currents.
c. mountain ranges.
d. deserts.

7. Most of the land area in the United States has a
a. polar climate.
b. tropical climate.
c. marine climate.
d. temperate climate.

8. Ice ages are thought to be associated with changes in
a. the Earth's axis and orbit.
b. ocean currents.
c. prevailing winds.
d. the sun's energy output.

True or False

If the statement is true, write "true." If it is false, change the underlined word or words to make the statement true.

1. The <u>temperate</u> zone is also called a high-latitude climate.

2. The <u>leeward</u> side of a mountain usually receives a lot of precipitation.

3. An area near the equator <u>does not</u> receive the direct rays of the sun.

4. As the amount of carbon dioxide in the atmosphere increases, the Earth's temperature will probably <u>increase</u>.

5. A land area near a large lake would probably have a <u>continental</u> climate.

6. The Gulf Stream, which flows away from the equator, is a <u>cold</u> water current.

7. A region that receives <u>less</u> than 25 centimeters of rainfall a year is a desert.

8. The warm periods between ice ages are called <u>major glaciations</u>.

Concept Mapping

Complete the following concept map for Section 2–1. Refer to pages K6–K7 to construct a concept map for the entire chapter.

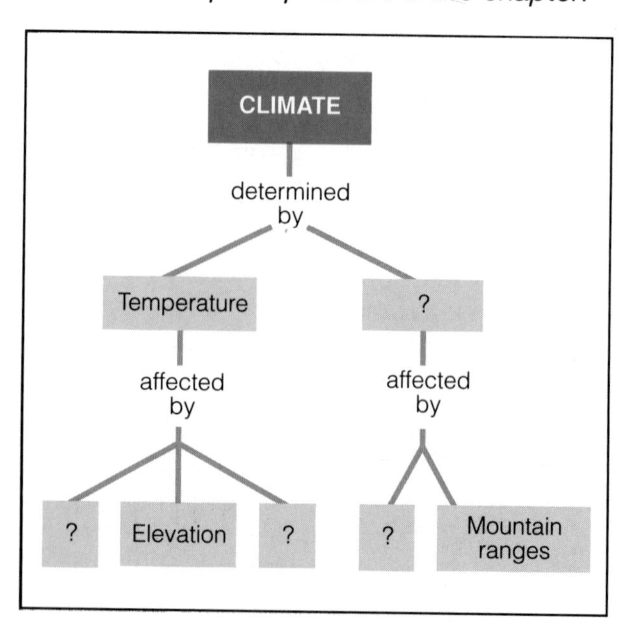

Concept Mastery

Discuss each of the following in a brief paragraph.

1. Describe how latitude, elevation, ocean currents, prevailing winds, and mountain ranges affect climate.
2. Explain how a large body of water such as an ocean can keep a nearby land area cool in the summer and warm in the winter.
3. Why do deserts have such a large daily temperature range? What are two mistaken beliefs that many people have about conditions in a desert?
4. How might changes in climate have resulted in the extinction of the dinosaurs?
5. Explain why the Sahara is one of the driest places on Earth even though it is near the Atlantic Ocean.
6. How was the climate during a major glaciation different from the climate during an interglacial? How was a major glaciation different from the climate today? What was the Little Ice Age?
7. Would a desert in the tropical zone be more likely to be found on the eastern or western coast of a continent? Why?

Critical Thinking and Problem Solving

Use the skills you have developed in this chapter to answer each of the following.

1. **Making predictions** Suppose a warm ocean current flowing along a coast suddenly becomes a cold ocean current. Predict what would happen to the climate along the coast. Explain why.
2. **Relating cause and effect** Explain how the tilt of the Earth's axis causes different seasons in the Northern Hemisphere and the Southern Hemisphere at the same time of the year.
3. **Interpreting a diagram** Study the diagram shown here. Which side of the mountain is the windward side? Which side is the leeward side? On which side would you be more likely to find a desert? Explain.

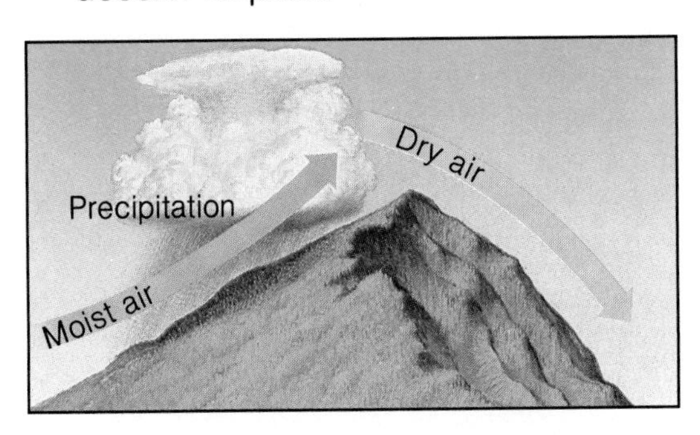

4. **Making inferences** Why do you think meteorologists record both the maximum and the minimum temperatures rather than the average temperature in order to determine the climate of a desert?
5. **Making inferences** Why do you think polar, temperate, and tropical climate zones are also called high-, middle-, and low-latitude climates, respectively?
6. **Relating concepts** The specific heat of a substance is the amount of heat needed to raise the temperature of 1 gram of the substance 1°C. Water has a relatively high specific heat. Use this information to explain why the presence of a large body of water has a moderating effect on temperatures in marine climate regions.
7. **Using the writing process** Write a science fiction story describing a trip by time machine into the future or the past. Be sure to include a detailed description of the climate in your story.

Climate in the United States

Guide for Reading

After you read the following sections, you will be able to

3–1 Climate Regions of the United States

- Describe the kinds of climates that occur in the United States.

3–2 Land Biomes of the United States

- Explain what a biome is and how land biomes are classified.
- Describe the kinds of land biomes located in the United States.

Herds of caribou were once common in Maine. Then hunting and other human activities drove the caribou farther north. Many people hope that Maine may once again be home to the caribou.

As part of the Maine Caribou Reintroduction Project, game wardens captured 30 caribou in Canada and brought them to the University of Maine at Orono. These 30 caribou became a "nursery herd" for breeding more caribou that were slowly introduced to their new homes in Maine. Members of the Reintroduction Project hoped that as many as 100 caribou would be roaming through Maine by 1996. Unfortunately, all of the caribou that were released mysteriously vanished. However, other programs to reintroduce deer herds in Pennsylvania and red wolves in New York were more successful. Perhaps the caribou will return to Maine in the future.

Today, caribou spend the summer in an area called the tundra. In the winter, the caribou often travel south to coniferous forests. In this chapter you will learn the difference between tundras and coniferous forests. You will also learn about the climate in these and other areas of the United States, and about the plants and animals that live there.

Journal *Activity*

You and Your World What kinds of plants and animals live in your part of the United States? In your journal, sketch two kinds of plants and two kinds of animals that are common where you live. Did you ever wonder why these plants and animals are characteristic of your area?

A solitary caribou bull will join with thousands of other caribou in vast herds to migrate in search of food.

Guide for Reading

Focus on this question as you read.

- *What are the six major climate regions of the United States?*

3–1 Climate Regions of the United States

Where in the United States do you live? Do you live among the corn and wheat fields of the Midwest plains? Do you live out on the West Coast? Do you live in the hot, dry deserts of the Southwest? Or are you one of the millions of people living along the Eastern Seaboard? Wherever you live, you know that your section of the United States has its own climate (the average weather conditions there over a long period of time).

The three climate zones that you learned about in Chapter 2—polar, temperate, and tropical—are all represented in various sections of the United States. Alaska is located in the polar zone. Hawaii and southern Florida are located in the tropical zone. But most of the United States is located in the temperate zone.

Within the temperate zone, there are many different climates. To describe each climate precisely, scientists have divided the mainland United States into six major climate regions. **The six climate regions of the United States are Mediterranean, marine west coast, moist continental, moist subtropical, desert, and steppe.** The division of the mainland United States into six major climate

Figure 3–1 *From the busy streets of New York City to an Iowa corn field to a tranquil rice paddy in Arkansas, each region of the United States has its own climate.*

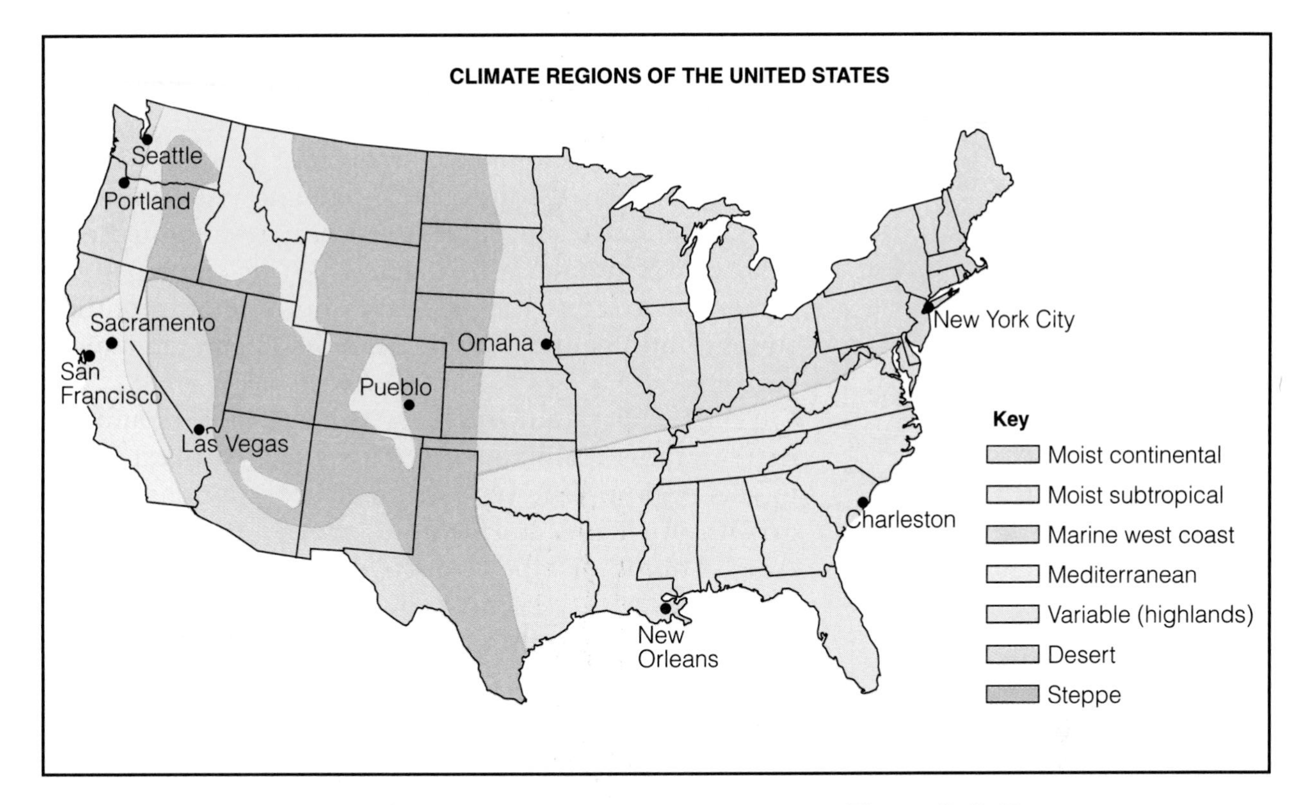

Figure 3–2 *The major climate regions of the United States are shown on this map, along with representative cities in each region. Remember that the climate does not change suddenly as you cross an invisible line from one climate region to another! The climate regions actually blend gradually into one another.*

regions is based on the average precipitation and temperature in each region. But it is important to remember that even within a particular climate region, variations in precipitation and temperature exist. The map in Figure 3–2 shows the six major climate regions of the United States. According to Figure 3–2, in which climate region do you live?

In the pages that follow, you will read about the six major climate regions of the United States. You will learn about the average weather conditions for each region, and also about some of the living things that are characteristic of the region. As you read this information, keep in mind the ways in which the climate of a particular region affects the plants and animals (including humans) that live in the region. Look for connections among the region's climate, ecology (relationship of living things to the environment), and economy (what people do for a living).

Mediterranean Climate Region

The narrow coastal area of California has a **Mediterranean climate.** The name of this climate region comes from the area around the Mediterranean Sea. In winter, cyclones and moist maritime

MEDITERRANEAN CLIMATE		
	Summer	**Winter**
Average Temperature (°C)		
San Francisco	17.3	10.0
Sacramento	24.0	9.0
Average Precipitation (cm/month)		
San Francisco	0.2	9.8
Sacramento	0.2	8.0

Figure 3–3 *Variations in temperature and precipitation occur within each climate region. San Francisco, California, and Sacramento, California, both have a Mediterranean climate. How did this climate region get its name?*

polar air masses bring heavy precipitation to the Mediterranean climate region. In summer, however, there is almost no rain. (This type of climate is also known as a dry-summer subtropical climate.) Winters throughout the Mediterranean climate region are cool. Summer temperatures are only slightly higher than winter temperatures. The chart in Figure 3–3 shows the average temperature and precipitation in summer and winter for the Mediterranean climate region.

The Mediterranean climate of wet winters and dry summers results in two basic types of plant life in this region. One type of plant life is a dense growth of shrubs and stunted (shorter than normal) trees. The other type consists of scattered oak and olive trees with a ground cover of grasses. You will learn more about the types of plants characteristic of each region of the United States in the next section.

With the help of extensive irrigation, agriculture is a major occupation in the Mediterranean climate region. Most crops produced in California are grown in the Central Valley. There are still some citrus groves near Los Angeles despite the city's expansion. Grapes are grown in many places, including the Napa Valley (the wine-making center) and Fresno (the raisin center). Other important crops include peaches, a variety of vegetables, rice, cotton, and alfalfa.

Figure 3–4 *Mixed shrubs and stunted trees are characteristic of California's Mediterranean climate. The Napa Valley in California is ideal for vineyards.*

Marine West Coast Climate Region

The northwestern coast of the United States has a **marine west coast climate.** This is a rainy climate because moist air from the Pacific Ocean rises, cools, and releases precipitation onto the western slopes of the Cascade Mountains. Mild winters and cool summers are characteristic of the marine west coast climate. The temperature range from one season to another is small due to the moderating effect of the nearby Pacific Ocean.

The type of plant life most common to the marine west coast climate region is the forests of needle-leaved trees. These thick forests consist of mixed cedar, spruce, redwood, and fir trees. In fact, the most valuable needle-leaved forests in the world are located in the Pacific Northwest. These forests contribute greatly to the economy of this region. Processing wood for lumber, paper, and furniture is the major industry.

MARINE WEST COAST CLIMATE

	Summer	Winter
Average Temperature (°C)		
Seattle	17.0	5.3
Portland	18.5	5.0
Average Precipitation (cm/month)		
Seattle	2.3	11.9
Portland	2.3	15.5

Figure 3–5 *Seattle, Washington, and Portland, Oregon, both have a marine west coast climate. What variations in temperature occur from summer to winter in this climate region?*

Moist Continental Climate Region

The northern portion of the United States extending from the Midwest (central Nebraska and Kansas) to the Atlantic coast has a **moist continental climate.** Continental polar air masses flowing south across this region produce very cold winters. In

Figure 3–6 *Mt. Hood in Oregon provides a dramatic background for the thick forests of needle-leaved trees characteristic of a marine west coast climate. These forests have made the lumber industry an important part of the economy.*

MOIST CONTINENTAL CLIMATE		
	Summer	**Winter**
Average Temperature (°C)		
New York City	22.0	–0.3
Omaha	23.7	–4.3
Average Precipitation (cm/month)		
New York City	10.0	8.4
Omaha	8.7	2.2

Figure 3–7 *New York City, New York, and Omaha, Nebraska, share a moist continental climate. What characteristics does this climate have?*

summer, tropical air masses flowing north across the region produce high temperatures throughout the region.

The moist continental climate region receives a moderate amount of precipitation throughout the year. During the summer, however, there is a marked increase in precipitation for all locations within the region. During the winter, the decrease in precipitation varies from place to place. Look at Figure 3–7. What is the difference in average monthly precipitation for Omaha, Nebraska, from summer to winter?

In some sections of the moist continental climate region, forests of broad-leaved and needle-leaved trees are dominant. In other sections, much of the land was once covered with tall grasses. Farms now have largely replaced both the forests and the tall grasses.

Commercial agriculture is the principle occupation in this region. The moist continental climate region includes all of the Corn Belt as well as the eastern half of the Winter Wheat Belt. After corn and wheat, soybeans are the next most important crop in the region. Hogs, poultry, and beef cattle are also a major part of the economy.

Figure 3–8 *The moist continental climate region includes both pine trees in Maine and the endless wheat fields of Colorado.*

PROBLEM Solving

What Causes a Drought?

In 1988, the midwestern United States experienced a severe drought—a prolonged period of extremely dry weather. Little rain fell to help irrigate the corn and wheat fields. Many farmers lost their entire crops, leading to great economic hardships in this region. What were the causes of this harmful drought?

The map shown here illustrates the events that led to the drought in the Midwest. The numbers show the order in which the events occurred. Study the map and then answer the questions that follow.

Relating Cause and Effect

1. What event began the process that led to the drought?

2. What was the final event that caused the drought to occur? What was the immediate cause of this event?

3. What unusual storm activity contributed to the drought? What was the direct result of this storm activity?

4. Summarize the ways in which changes in global winds and changes in the temperature of ocean water contributed to the drought.

5. How are the changes you described in question 4 related?

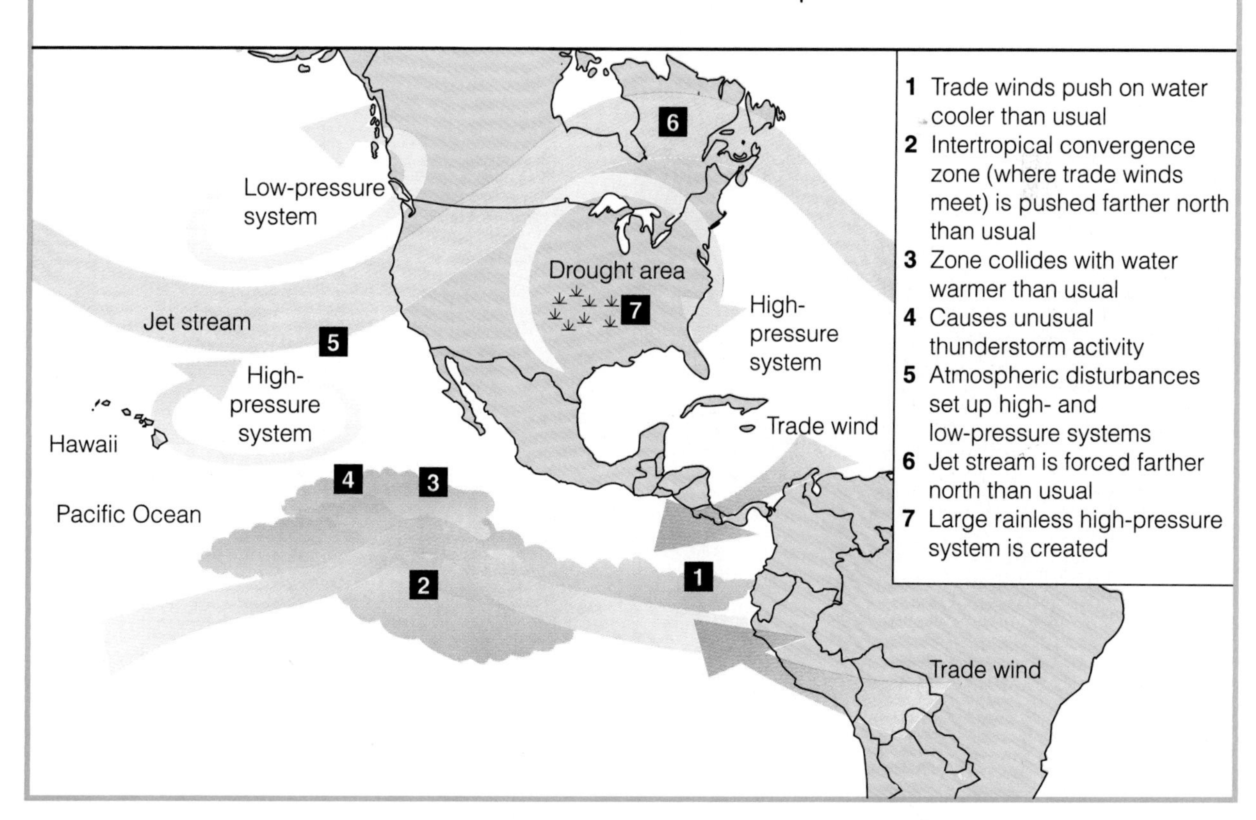

MOIST SUBTROPICAL CLIMATE		
	Summer	**Winter**
Average Temperature (°C)		
New Orleans	26.3	12.7
Charleston	27.2	11.3
Average Precipitation (cm/month)		
New Orleans	14.0	10.1
Charleston	15.0	7.6

Figure 3–9 *A moist subtropical climate is typical of New Orleans, Louisiana, and Charleston, South Carolina. Are the variations in temperature and precipitation from summer to winter extreme in this climate region?*

Moist Subtropical Climate Region

The southeastern United States has a **moist subtropical climate.** Summers are hot in the moist subtropical climate region. The average precipitation in summer is greater than it is in winter. In fact, the characteristic summer temperatures and precipitation in the moist subtropical region are similar to those of the tropical climate zone. Maritime tropical air masses moving inland from the tropical zones greatly influence the summer climate in this region. So in summer, the climate of the moist subtropical region of the United States is similar to the climate of the Earth's tropical zones. (As you may recall from Chapter 2, the tropical zones extend from 30° north and south latitude to the equator.)

The similarity between the moist subtropical climate and the tropical climate ends in winter. Although winters in the moist subtropical climate region are generally cool and mild, the mixing of polar air masses with maritime tropical air masses causes the temperature to drop below freezing occasionally. Severe frosts sometimes occur in the northern areas of the region. In late summer and early autumn, hurricanes are common along the coast. In spring and summer, tornadoes are common in the western parts of the region.

Figure 3–10 *Live oak trees flourish in Florida, which has a moist subtropical climate. Cotton, which was once the major crop in the South, is still grown today in many southern states.*

Figure 3–11 *The American alligator and the green heron are two inhabitants of the Florida Everglades.*

The plant life of the moist subtropical climate region consists of forests of broad-leaved and needle-leaved trees. Oak, chestnut, and pine trees grow in this region. The "River of Grass" of the Florida Everglades is also located in this region. The Everglades is home to a large variety of animals, including herons, egrets, alligators, crocodiles, and manatees.

At one time, cotton was the major crop in the South. Cotton is still grown in the region today, but other crops have become an important part of the economy. For example, citrus fruits are grown in central Florida, peaches and peanuts in Georgia, sugar cane in Mississippi, and rice in Louisiana.

Desert and Steppe Climate Regions

Located within the western interior of the United States are two regions that have similar climates.

Life in the Everglades

Marjorie Kinnan Rawlings (1896–1953) wrote many novels concerning life in the Florida Everglades, including the classic *The Yearling.* Her essay "Rattlesnake Hunt" vividly describes the desolate region in Florida where rattlesnakes live.

DESERT AND STEPPE CLIMATES		
	Summer	Winter
Average Temperature (°C)		
Las Vegas	28.3	8.2
Pueblo	22.5	-0.3
Average Precipitation (cm/month)		
Las Vegas	10.3	1.4
Pueblo	4.2	1.2

Figure 3–12 *The desert and steppe climates of Las Vegas, Nevada, and Pueblo, Colorado, receive the lowest amounts of precipitation of any region in the United States.*

These climates are the **desert climate** and the **steppe climate.** The desert climate region and the steppe climate region begin just east of the mountain ranges along the west coast and end in the central midwestern part of the United States (the Great Plains).

The desert and steppe climate regions receive the lowest amount of precipitation of any climate region in the United States. However, the steppe climate region receives slightly more precipitation than the desert climate region does.

One reason precipitation is so low in these climate regions is that they are located far inland, away from the oceans that are the sources of moist maritime air masses. (The desert and steppe climate regions are also called dry continental climates.) In addition, high mountain ranges along the western borders of these regions block most of the maritime air masses. In winter, dry continental polar air masses further reduce the amount of precipitation these regions receive.

In spite of the harsh conditions found in the desert climate region, many plants—including cactus, yucca, and sagebrush—grow well in this region. The slightly higher precipitation of the steppe climate encourages the growth of short grasses and scattered forests of needle-leaved trees.

Figure 3–13 *Several types of cacti and other hardy plants are able to survive the harsh conditions of the southwestern deserts. Sheep herding is important to the economy of the steppe climate region.*

Grazing for livestock—beef cattle and sheep—is an important part of the economy of the Great Plains. Goats are also raised in this region, especially in Texas. Because this region is so dry, overgrazing can be a serious problem. Overgrazing exposes the dry soil so that it can be blown away by the wind. These conditions led to the dust bowl of the 1930s.

Highlands (Variable) Climate Region

You probably noticed a climate region identified as highlands, or variable, in Figure 3–2 on page 85. These regions are located in mountain areas. The climate varies with latitude and elevation. Generally, the temperature in highlands regions is low—the higher the elevation, the lower the temperature. Average precipitation increases with elevation and is higher than in the surrounding lowlands.

What's in a Name?

Grasslands in different parts of the world are known by many names. Using library references, find out where each of the different types of grasslands listed here is found. Write a brief report describing the location and characteristics of each of the following grassland biomes: steppes, savannas, pampas, plains, prairies, llanos, veldts.

Figure 3–14 *This diagram illustrates the effects of latitude and elevation on climate. Climbing 1000 meters higher in elevation has about the same effect as traveling 1700 kilometers north. How are conditions at the highest elevations on a mountain similar to those at the North Pole?*

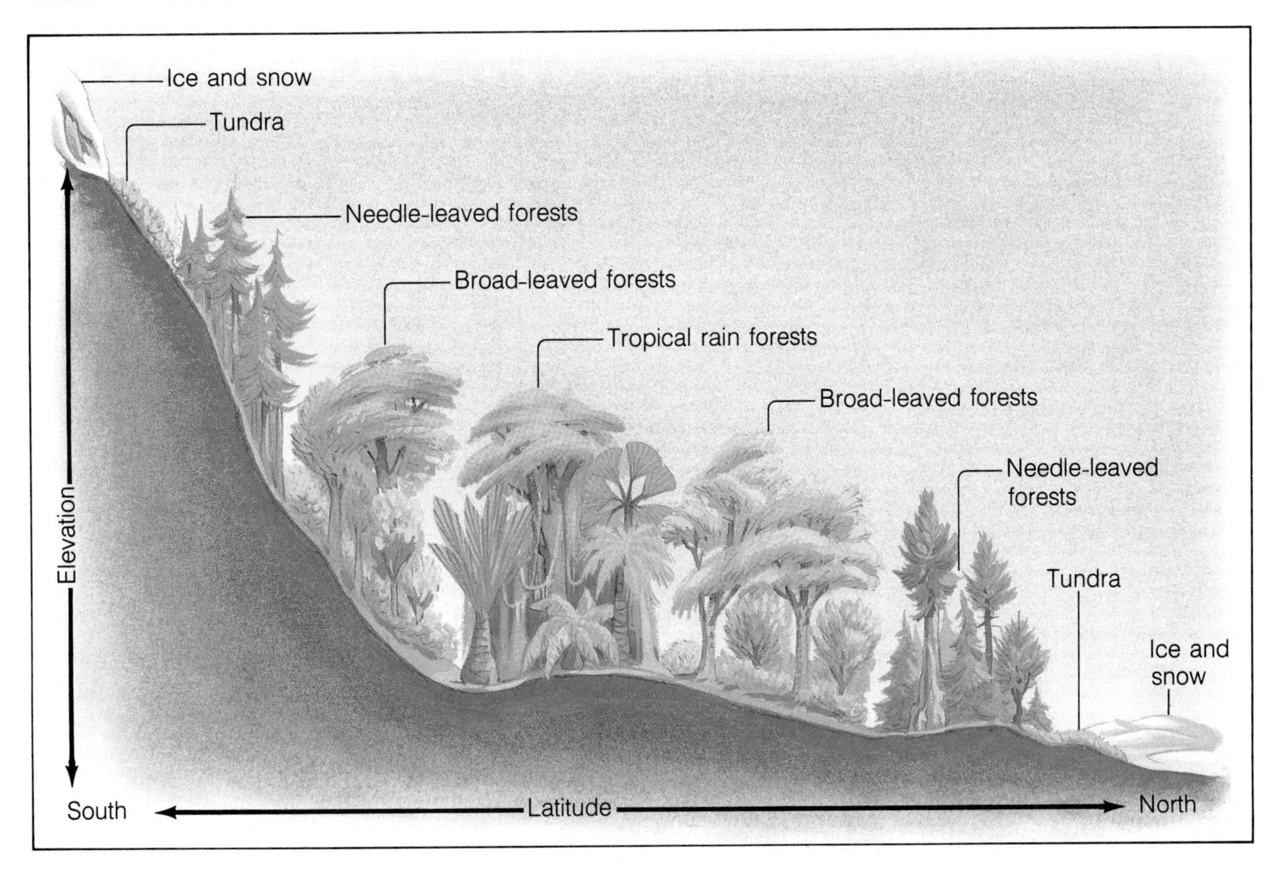

Forestry is the major industry in the mountain regions of the western United States. Fir and pine are the most important commercial trees. Mountain areas are also used as summer grazing grounds for livestock (especially sheep), which are transported from the Great Plains.

3–1 Section Review

1. What are the major climate regions of the United States? What factors determine this classification?
2. Why are summer temperatures in San Francisco, California, cooler than summer temperatures in Sacramento, California?
3. What are three reasons for the low precipitation in desert and steppe climate regions?
4. Why does the east coast of the United States receive more precipitation in the summer, whereas the west coast receives more precipitation in the winter?

Critical Thinking—*Making Comparisons*

5. What changes in climate would you expect if you were to travel from New York City to New Orleans, Louisiana, during the summer? What if you were to make the same trip during the winter?

Guide for Reading

Focus on this question as you read.

- *What are the major land biomes of the United States?*

3–2 Land Biomes of the United States

Imagine that you are taking a trip across the United States. As you travel from one part of the country to another, you quickly discover that different plants and animals live in different areas. So you would probably be really surprised to see an alligator in the middle of the Mojave Desert or a cactus in the Everglades!

Why do some kinds of plants and animals survive in one area but not in another? Why do different groups of plants and animals live in different areas?

You can probably answer these questions easily once you think about them. The kinds of animals that live in an area depend largely on the kinds of plants that grow there. For example, the midwestern United States was once home to millions of grazing bison. Today, however, most of the grasses that once grew in the Midwest have been replaced by corn and wheat fields. What effect do you think this has had on the bison population? You are right if you said that the herds of bison have greatly decreased and have almost disappeared. (Today, small herds of bison have been established on private land.)

If the plant life in an area determines the animal life in that area, what determines the plant life? The plant life in an area is determined mainly by climate. Recall that climate refers to the general conditions of temperature and precipitation for an area over a long period of time. As you learned in Section 3–1, the United States has six major climate regions. Scientists classify areas with similar climates, plants, and animals into divisions called **biomes** (BIGH-ohmz).

Biomes are divisions that help scientists better understand the natural world. But as is often the case in science, not all scientists agree on the kinds and number of biomes. However, most scientists accept at least six land biomes. Each of these biomes is located in some area of the United States. **The major land biomes of the United States are tundras, coniferous forests, deciduous forests, tropical rain forests, grasslands, and deserts.** There are also several types of aquatic, or water, biomes. Because aquatic biomes do not depend on climate, they will not be considered here.

Figure 3–15 *At one time, millions of bison could be seen grazing on the Great Plains. Today, only a few scattered herds remain.*

Microbiomes

Select a 2-meter square portion of land in your neighborhood. Visit your "microbiome" once a week for several weeks and observe the kinds of organisms that live there. Record your observations in a notebook. Notice how and when organisms move into and out of the area, as well as the weather conditions each day.

■ What relationships exist in your microbiome? Explain.

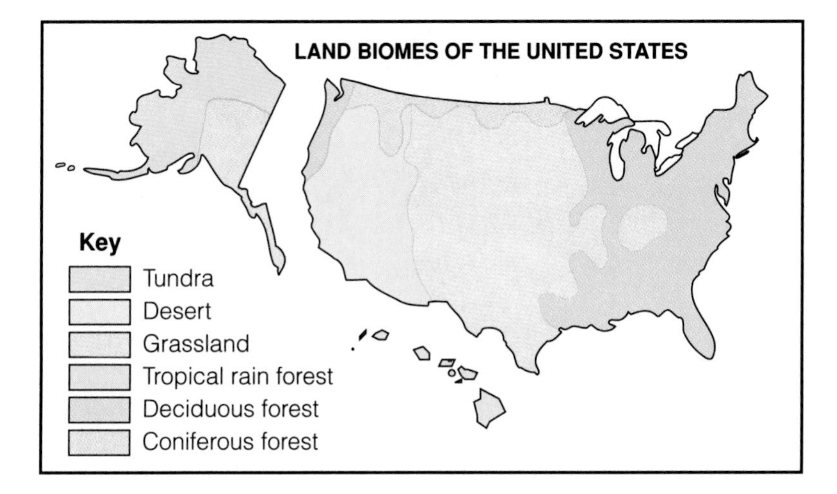

Figure 3–16 *This map shows the distribution of major land biomes in the United States. In which biome do you live?*

Tundras

Figure 3–17 *British soldier lichen (top) are common on the tundra as are tundra willow, dwarf birch, and sedge (bottom).*

Tundra biomes cover about 10 percent of the Earth's surface. In the United States, tundra biomes are found only in parts of Alaska. (Recall from Chapter 2 that Alaska is in the polar climate zone.) The climate of a tundra is extremely cold and dry. In fact, you could think of a tundra as a cold desert. Less than 25 centimeters of rain and snow fall on a tundra during most years. (On the average, a snowfall 10 centimeters deep is the equivalent of 1 centimeter of rainfall.) The little water that is found on a tundra is permanently frozen in the soil. This frozen layer of soil is called **permafrost.** Almost 85 percent of the ground in Alaska is permafrost.

Plant life on a tundra consists mostly of mosses and grasses. Carpetlike lichens, which are actually fungi and algae growing together, cover the rocks and bare ground. Because of the permafrost, large trees cannot root on a tundra. The few trees that do grow here are mainly knee-high willows and birches.

Lichens are the favorite food of the caribou herds you read about at the beginning of this chapter. The caribou roam the tundras in the summer before moving farther south for the winter. Wolves often follow close behind the caribou, preying on the old and weak animals. Birds such as ptarmigan and small animals such as lemmings also inhabit the tundras. Some animals are only seasonal residents of the tundras. Arctic terns, for example, make round-trip migrations of 34,000 kilometers to mate and raise their young during the short tundra "summer."

Figure 3–18 *The Alaska tundra is home to the white-tailed ptarmigan, the arctic tern, and the shaggy musk ox.*

Coniferous Forests

South of the tundra biomes are the coniferous forest biomes. Unlike the permafrost of the tundras, the soil in a coniferous forest thaws every spring, making the forest floor wet and swampy. For this reason, a coniferous forest biome is also called a **taiga** (TIGH-guh), a Russian name that means swamp forest. A coniferous forest biome, or taiga, is found in parts of Alaska as well as at the higher elevations of the Rocky Mountains. Temperatures in coniferous forest biomes are cold. The yearly rainfall is between 50 and 125 centimeters.

Few types of trees can survive the cold climate of the coniferous forests. The trees that do live in these biomes are needle-leaved trees, or **conifers.** Conifers produce their seeds in cones. They include firs, spruces, and pines. Giant redwoods grow along the coasts of Washington State, Oregon, and northern California. These conifers, which may grow as tall as 60 meters, are among the tallest trees in the world. (The tallest redwood ever found was 110 meters tall—almost 20 meters taller than the Statue of Liberty!) The Mediterranean climate of southern California supports a coniferous forestlike biome called a chaparral. A chaparral consists mainly of short, shrublike plants.

Large animals in the coniferous forests include wolves, deer, black bears and grizzly bears, and moose. (Parts of the coniferous forests are even called ''spruce-moose'' belts.) Many smaller animals, such as beaver, hares, and red squirrels, also live in

Figure 3–19 *Soaring redwoods and other conifers can be seen in Sequoia National Park, California.*

Activity Bank

Soil Permeability, p.126

Figure 3–20 *The brown bear shares its home in the coniferous forest with the moose and the great horned owl. What is another name for a coniferous forest?*

Figure 3–21 *In autumn, a deciduous forest becomes a blaze of vivid colors. When the leaves decay, nutrients that help new trees grow are returned to the soil.*

the coniferous forests. Crows and great horned owls are some of the birds that build their nests among the conifers. Grouse roost in the branches.

Deciduous Forests

South of the coniferous forest biomes are the deciduous forests. Deciduous forests begin at the northeastern border, between the United States and Canada, and cover the eastern United States. Deciduous trees shed their leaves in the autumn. New leaves grow back in the spring. The summers in the deciduous forests are warm and the winters are cold, but they are not as cold as in the coniferous forests. Rainfall in the deciduous forests is between 75 and 150 centimeters a year.

There are more than 2500 kinds of deciduous trees. Oak, birch, maple, beech, and hickory are the most common varieties found in the deciduous forests of the United States. Autumn in the deciduous forests is one of the most beautiful seasons of the year because of the bright colors the leaves display before they fall to the ground. In the spring, wildflowers and ferns cover the forest floor.

Many different kinds of animals make their homes in the deciduous forests. Thrushes, woodpeckers, cardinals, and blue jays are some of the many birds you might see in a deciduous forest. Snails, worms, snakes, and salamanders slither along the forest floor. Small mammals, such as gray squirrels and raccoons, live among the branches of the trees.

Figure 3–22 *The hollow trunk of a deciduous tree makes a cozy home for the pileated woodpecker and its young, as well as for these baby raccoons. The black bear cub is just visiting, however! Black bears live in dens on the ground.*

Tropical Rain Forests

In the United States, tropical rain forests are found only in Hawaii. (Recall that Hawaii is in the tropical climate zone.) As you might expect, rain forests get a great deal of rain—at least 200 centimeters a year. Kauai, Hawaii, may be the wettest place on Earth. It receives an average rainfall of 1215 centimeters every year! Temperatures in the tropical rain forests remain warm all year, so plants grow well here throughout the year.

Rain forests have more varieties of plant life than any other biome. Trees grow to a height of 35 meters or more. High above the forest floor, the tops of the trees meet to form a green roof, or layer, called a **canopy** (KAN-uh-pee). The canopy is so dense that rainfall may not reach the forest floor for 10 minutes after hitting the canopy! Most of the other plants in a rain forest grow in the canopy, where sunlight can reach them. Woody vines up to 90 meters long hang from the trees. Orchids and ferns grow on the branches of trees instead of on the ground.

Like the plant life, animal life in a rain forest is rich and varied. Some rain forest animals spend their entire lives high in the trees and never touch the ground. Parrots, toucans, and hundreds of other birds live in the canopy. At night, huge colonies of bats come out to hunt among the trees. Insects, tree frogs, and snakes crawl on the trunks and branches of the trees.

Daily Rainfall

The island of Kauai, Hawaii, receives an average of 1215 centimeters of rain every year. What is the average rainfall each day?

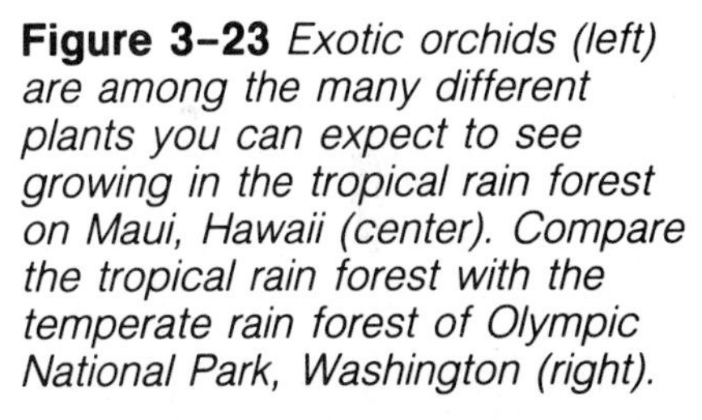

Figure 3–23 *Exotic orchids (left) are among the many different plants you can expect to see growing in the tropical rain forest on Maui, Hawaii (center). Compare the tropical rain forest with the temperate rain forest of Olympic National Park, Washington (right).*

Figure 3–24 *The iiwi and the nene goose are birds that are native to Hawaii. The ti plant grows on the slopes of Hawaiian volcanoes.*

Grasslands

The first European explorers found an endless sea of grass in the midwestern plains of the United States. French explorers from Canada called these grasslands a prairie, a French word that means meadow. Some of the grasses on the prairie were over 2 meters tall! The grassland biomes in the United States receive between 25 and 75 centimeters of rain every year. The grasslands of the midwestern plains are characterized by hot summers and cold winters.

Grasses make up the main group of plants in a grassland biome. There are few trees because of the low rainfall. Fires, which often sweep over the grasslands, also prevent widespread tree growth. Today, most of the original grasslands in the United States have been replaced by farms and pastures. Wheat, corn, and other grains are now widely farmed in the midwestern plains of the United States.

Gophers, prairie dogs, and other small animals live on the grasslands. Blackbirds, prairie chickens, and meadow larks are among the birds that feed on the grasshoppers, locusts, and other insects. Large plant eaters, such as elk and bison, were once common on the plains. They were hunted by wolves and cougars. Now that farms have replaced most of the original grasslands, however, most of the large animals live only in national parks and other protected areas.

Figure 3–25 *Although most of the midwestern plains have been converted to corn or wheat fields, native grasslands have been restored in the Flint Hills Preserve in Kansas.*

CAREERS

Biogeographer

No matter where in the United States you live, you are probably familiar with the plants and animals that are common in your region. For example, you would see palm trees in Florida, cacti in Arizona, and redwoods in California. Scientists who study the distribution of plants and animals throughout the world are called **biogeographers.**

Biogeographers must know about botany, zoology, meteorology, and politics. If you enjoy reading, doing research, and working outdoors, you might want to consider biogeography as a career. To learn more about this field, write to the Association of American Geographers, Biogeography Specialty Group, 1710 16th Street, NW, Washington, DC 20009.

Figure 3–26 *The greater prairie chicken is a common resident of the grasslands. The round structure in its throat is its vocal sac, not an orange! Two other inhabitants are the grasshopper and the tiny harvest mouse.*

ACTIVITY

DISCOVERING

Plant Adaptations

Obtain a small cactus plant. Using a hand lens, carefully observe the outside of the cactus. What do you think the inside of the cactus will look like? Why?

Use a knife to cut off the tip of the cactus. **CAUTION:** *Be careful when using a knife.*

How did the inside of the cactus compare with your predictions?

■ How do the adaptations of the cactus help it survive?

Deserts

Unlike the other biomes, deserts can be classified by what they do not have: water. Deserts receive less than 25 centimeters of rain a year. Desert biomes are located in the southwestern part of the United States. Although deserts can be hot or cold, the deserts of the American Southwest are hot.

Plants in a desert are adapted to the lack of rainfall. For example, the thick, fleshy stems of cacti help them to store water. A giant saguaro cactus in the Sonoran Desert of Arizona can store up to a ton of water! The Joshua tree, a giant yucca, is one of the few trees that can survive in the deserts of the Southwest. Most flowering plants in the southwestern deserts flower, produce seeds, and die within a few weeks of a rare desert rainfall.

Like the plants, desert animals must be able to survive on little water. Plant-eating animals, such as kangaroo rats and jack rabbits, obtain most of their water from the plants they eat. Meat-eating animals, such as cougars, obtain most of their water by eating

Figure 3–27 *Tall saguaro cacti may take hundreds of years to reach full height. Flowering hedgehog cacti and organ pipe cacti must bloom and produce seeds in a brief growing season. The Joshua tree is one of the few trees in the desert.*

Figure 3–28 *The jack rabbit, the mountain lion, and the white-winged dove all have adaptations that allow them to survive in the desert. What adaptations for survival does the cactus have?*

the plant eaters. Most desert animals hide from the hot sun during the day and come out to eat only at night, when temperatures are cooler.

3–2 Section Review

1. Identify and describe the main characteristics of the major land biomes of the United States.
2. Compare the climates of the three forest biomes.
3. What is another name for a coniferous forest?
4. How are plants and animals adapted for life in a desert?

Connection—*Social Studies*

5. About 10 percent of all workers in the seven states that make up the midwestern plains are employed on farms, whereas only about 1 percent are employed in forestry. Based on your knowledge of the midwestern plains biome, explain why this is so.

CONNECTIONS

A Pharmacy on the Prairie

Growing among the tall and short grasses of the prairie are many colorful wildflowers and weeds. The Native Americans who lived on the Great Plains before the coming of settlers from the East learned to use some of these plants as remedies, or cures, for common illnesses. This type of traditional *medicine* using plants is called folk medicine. Scientists have found that some traditional folk remedies have real medicinal value.

A common grasslands weed called fleabane was used by the people living on the prairie as an insect repellant. (The word bane means poison or killer. So fleabane is a ''flea killer.'') Fleabane was also used to cure sore throats and to help heal minor cuts and bruises. Laboratory tests have confirmed that substances in fleabane promote healing and help protect against infection.

An herb called blue cohosh was used by Native American women as an aid in childbirth. Before giving birth, pregnant women drank a tea made from the plant's roots. Modern research has shown that a substance derived from blue cohosh stimulates contractions of the uterus, or birth canal, which could lead to a faster and easier birth.

There are many other examples of folk remedies that have only recently been found to have a scientific basis. These examples should remind us to use caution when we change the *environment* to suit our own needs. By destroying wilderness areas, we may unknowingly be destroying many beneficial organisms at the same time.

Common fleabane

Blue cohosh

Laboratory Investigation

Comparing Climate Regions and Biomes

Problem

How can you use climate information to determine the biomes of the United States?

Materials *(per student)*

tracing paper
5 different-colored pencils

Procedure

1. Study the map of North America shown here. The lines on the map are isotherms (connecting places with the same temperature). The key shows the average yearly precipitation in different areas.
2. On a sheet of tracing paper, trace the outline of the United States from the map of North America. **Note:** *Be sure to trace Alaska.* (Hawaii, which is a tropical rain forest biome, is not shown on this map.)

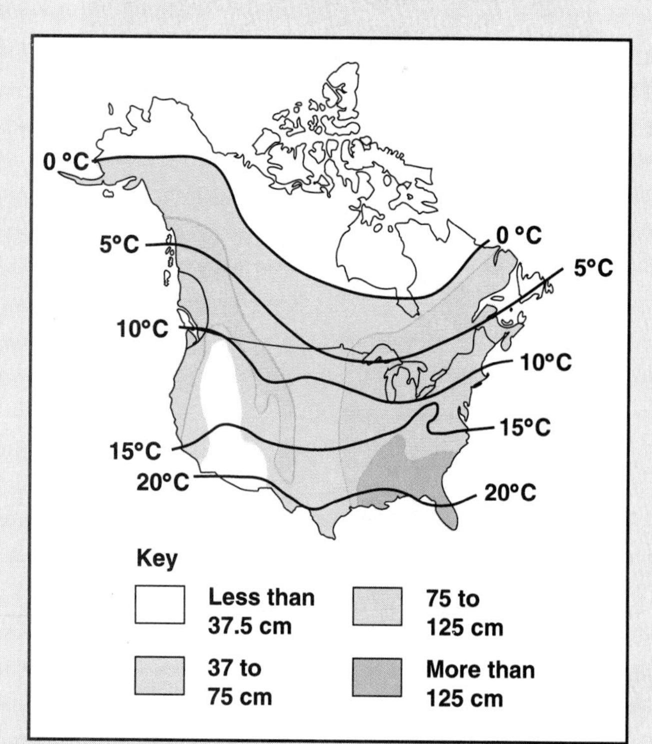

3. Study the data table, which shows the average yearly precipitation and temperature in five land biomes of the United States.
4. Using the information in the data table, draw the approximate boundaries of each biome on your map of the United States. **Note:** *Be sure to use a different-colored pencil for each biome.*

Observations

1. What is the average yearly temperature in each of the biomes listed in the data table?
2. What is the average yearly precipitation in each of the biomes listed in the data table?

Analysis and Conclusions

1. What happens to the average yearly precipitation as you travel across the United States from the west coast to the east coast?
2. Which biome covers the largest land area in the United States?
3. What is the average yearly temperature and precipitation where you live? In which biome do you live?
4. **On Your Own** Add a key to your map showing the main crops grown in each biome.

Biome	Average Yearly Precipitation	Average Yearly Temperature
Tundras	0 to 25 cm	below 0°C
Coniferous forests	50 to 125 cm	0° to 12°C
Deciduous forests	75 to 150 cm	5° to 25°C
Grasslands	25 to 75 cm	5° to 25°C
Deserts	0 to 25 cm	10° to 25°C

Summarizing Key Concepts

3–1 Climate Regions of the United States

▲ Most of the United States is located in the temperate climate zone.

▲ The six major climate regions of the United States are Mediterranean, marine west coast, moist continental, moist subtropical, desert, and steppe.

▲ The classification of climate regions is based on the average temperature and precipitation in each region.

▲ In the highlands, or variable, climate region, climate varies with latitude and elevation.

▲ The climate, ecology, and economy of a particular region are all interrelated.

3–2 Land Biomes of the United States

▲ The kinds of animals that live in an area depend on the kinds of plants that grow in the area, which are determined mainly by the climate of the area.

▲ Areas with similar climates, plants, and animals are classified into biomes.

▲ The major land biomes of the United States are tundras, coniferous forests, deciduous forests, tropical rain forests, grasslands, and deserts.

▲ Aquatic, or water, biomes are not determined by climate.

Reviewing Key Terms

Define each term in a complete sentence.

3–1 Climate Regions of the United States

Mediterranean climate
marine west coast climate
moist continental climate
moist subtropical climate
desert climate
steppe climate

3–2 Land Biomes of the United States

biome
permafrost
taiga
conifer
canopy

Chapter Review

Content Review

Multiple Choice

Choose the letter of the answer that best completes each statement.

1. The climate zone that includes most of the mainland United States is the
   a. polar zone.
   b. temperate zone.
   c. tropical zone.
   d. subtropical zone.
2. The type of climate found along the coast of California is called a
   a. marine west coast climate.
   b. moist continental climate.
   c. Mediterranean climate.
   d. moist subtropical climate.
3. Which of the following trees is a conifer?
   a. pine   c. beech
   b. oak   d. hickory
4. In the United States, the type of biome found only in Alaska is a
   a. desert.   c. coniferous forest.
   b. grassland.   d. tundra.
5. Redwoods, which are among the tallest trees in the world, grow in
   a. tropical rain forests.
   b. coniferous forests.
   c. deciduous forests.
   d. tundras.
6. The most important crops grown in the moist continental climate region of the United States are
   a. trees for lumber.
   b. fruits and vegetables.
   c. wheat and corn.
   d. cotton and citrus fruits.
7. The type of biome that has a greater variety of plant life than any other biome is the
   a. coniferous forest biome.
   b. deciduous forest biome.
   c. grassland biome.
   d. tropical rain forest biome.

True or False

If the statement is true, write "true." If it is false, change the underlined word or words to make the statement true.

1. Dry-summer subtropical climate is another name for the marine west coast climate region.
2. The Russian name for a deciduous forest is taiga.
3. The climate in a variable, or lowlands, climate region varies with latitude and elevation.
4. Climate is not a factor in determining water biomes.
5. Like other coniferous trees, oak and maple trees shed their leaves in autumn.
6. The Florida Everglades is located in the moist continental climate region.

Concept Mapping

Complete the following concept map for Section 3–1. Refer to pages K6–K7 to construct a concept map for the entire chapter.

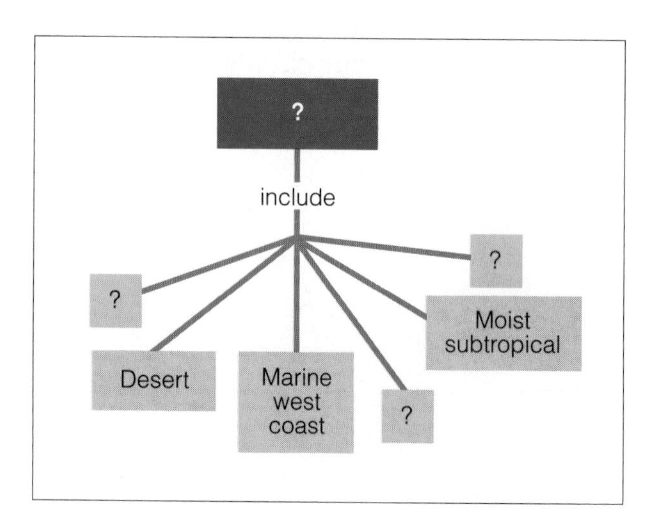

Concept Mastery

Discuss each of the following in a brief paragraph.

1. Briefly describe the kinds of plants and animals that live in each of the six major land biomes of the United States.
2. Why are tundra biomes sometimes referred to as cold deserts?
3. What crops are grown in the Mediterranean climate region? Why is extensive irrigation needed to grow crops in this region?
4. Why is climate an important factor in determining land biomes?
5. Why do you think it is difficult for scientists to agree on the number and kinds of biomes?
6. How are the desert and steppe climate regions similar? How are they different?
7. Why is wood processing a major industry in the Pacific Northwest?
8. How is the climate of the moist subtropical climate region in the United States similar to the climate of the Earth's tropical climate zones? How is it different?

Critical Thinking and Problem Solving

Use the skills you have developed in this chapter to answer each of the following.

1. **Making charts** Make a chart in which you list each of the six major land biomes in the United States, the average yearly temperature and precipitation in each biome, and the kinds of plants and animals that live in each biome.
2. **Classifying** In which climate region or regions of the United States would you expect to find each of the following plants?
   a. olive trees
   b. tall grasses
   c. redwood trees
   d. sagebrush
   e. pine trees
   f. short grasses
3. **Classifying** In which biome or biomes would you expect to find each of the following animals?
   a. caribou
   b. parrots
   c. raccoons
   d. moose
   e. kangaroo rats
   f. prairie dogs
4. **Applying concepts** As you climb a mountain, you may pass through several biomes. Explain how this is possible.
5. **Making predictions** The Arctic National Wildlife Refuge is part of the tundra biome in northeastern Alaska. The refuge is home to caribou, musk oxen, polar bears, and other animals. Now some oil companies want to explore the refuge for oil and natural gas. How might large-scale development of the refuge affect the animals (and plants) living in this biome?
6. **Interpreting photographs** Which biome is shown in this photograph? How do you know? What kinds of animals would you expect to find in this biome?

7. **Using the writing process** Pretend that you and your family are pioneers who have just settled on the Great Plains. Write a letter to a friend back home on the east coast describing life on the prairie during the 1800s.

JOANNE SIMPSON'S STORMY STRUGGLE

Joanne Simpson spent the year 1943 contributing to the American effort in World War II by teaching weather forecasting to air force personnel. At the same time, she was fighting a more personal battle. Simpson, who had graduated from the University of Chicago, wanted to return there to earn additional degrees in meteorology. Her efforts, however, were met with much opposition from professors at the university. They told Simpson that the idea of a woman meteorologist was a "lost cause." The concept of a female scientist, they claimed, was a "contradiction in terms," and "there was no point" in her trying to get an advanced degree. Yet, in 1949, the determined Simpson became the first American woman to receive a PhD in meteorology.

Simpson's interest in meteorology began when she was a young girl. Her father was

▶ From information gathered by weather satellites, meteorologists such as Joanne Simpson hope to learn more about various weather conditions and to be able to forecast severe storms.

a journalist who wrote about aviation, and she sometimes went flying with him. During her teen years, Simpson spent summers as an assistant to the director of aviation for the state of Massachusetts. She also began to take flying lessons.

In 1940, Simpson enrolled at the University of Chicago. She was introduced to meteorology while training for her pilot's license. The ability to read and understand a weather map, as well as knowledge of weather patterns and the atmosphere, are important to flying. Simpson was so fascinated by the subject that she signed up for a course at the university.

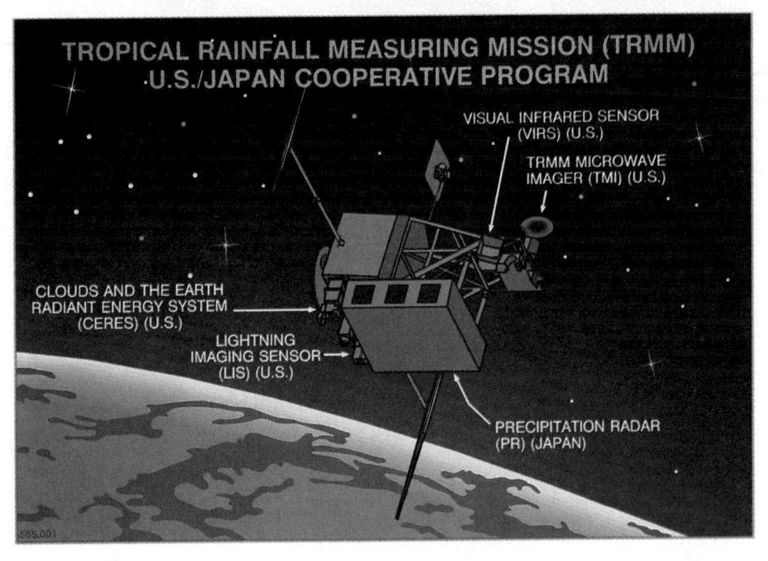

▲ **This diagram shows what the Tropical Rainfall Measuring Mission satellite will look like when it is placed in orbit around the Earth.**

After receiving her undergraduate degree and teaching military personnel for a year, Simpson decided to continue her studies in meteorology. At first none of the faculty at the University of Chicago would support her venture. But Simpson, a very determined young scientist, eventually won the support of Herbert Riehl. He agreed to supervise her research project, which involved the study of cumulus clouds—their interaction with the environment and their relationship to tropical waves.

Woods Hole, Massachusetts, a small town at the southwest tip of Cape Cod, provided an ideal natural environment for Simpson's research on cumulus clouds. Simpson studied at both Woods Hole and the University of Chicago before receiving her PhD in meteorology in 1949.

Simpson's interesting and dynamic career has included teaching at several universities—among them, the University of California at Los Angeles and the University of Virginia. In 1979, she was invited to head the Severe Storms Branch of the Goddard Space Flight Center Laboratory for Atmospheric Sciences. This laboratory is part of the National Aeronautics and Space Administration (NASA).

Dr. Simpson enjoys a challenge, so for her NASA is a perfect place to be. She continues to study the formation and development of cumulus clouds. And she is currently working on a new weather satellite that will provide accurate measurements of the rainfall in tropical ocean areas. The satellite, called TRMM (Tropical Rainfall Measuring Mission), which is scheduled to be launched in 1996 or 1997, should give meteorologists a better understanding of various changes in climate. These discoveries could enable scientists to make more accurate and longer range weather forecasts.

In the last 40 years, Simpson has published 115 scientific research papers, won numerous awards and honors, traveled across the globe, and served on many scientific councils and committees. The honors that have given her the greatest personal satisfaction are the Meisinger Award, given to her in 1962 for her work on cumulus clouds; NASA's Exceptional Scientific Achievement Medal; and the Carl-Gustav Rossby Research Medal, the American Meteorological Society's highest award. Simpson's courage and determination have earned her the respect of her colleagues and the public. She has truly paved the way for a generation of women meteorologists.

MAKING THE

Start with plenty of sunlight and good soil. Add lots of water. What is the result? Some of the world's richest farmland located in the western and southwestern United States. Today, much of this area is green and golden with fruits, vegetables, grain, and cotton. But some people remember when this land was dry, dusty desert and arid scrubland. How was a desert transformed into fertile farmland?

Irrigation has made the difference. Canals and pipelines now bring the water from distant lakes and rivers to desert regions that were once dry and parched. Wells dug deep into natural underground springs bring water to the surface. The water is then piped through fields and sprinkled on crops.

Some experts say that the price of growing crops in desert areas may be too high. Over the past 60 years, billions of dollars have been spent to bring water to dry desert areas. But the largest costs are not measured in money. Over time, irrigation causes serious problems.

One of the problems is the rapid depletion of general water reserves in order to provide water for irrigation. People are using water faster than nature can replace it. Reservoirs and wells that once held large amounts of water are now running dry.

Farmers in California, for example, rely on the water in 1200 reservoirs throughout the state to irrigate their crops. This water is carried by aqueducts and piped onto fields of cotton, alfalfa, and rice. Basically, what

▼ **In 1915, the Imperial Valley in California was dry and barren. Today, irrigation of this valley and other desert areas provides good land for crops to grow.**

DESERT BLOOM

IS THE PRICE TOO HIGH?

▲ **Thanks to irrigation canals, the Imperial Valley has become fertile farmland. But is the price too high?**

the farmers are doing is growing monsoon-climate crops in the middle of a desert! By 1991, however, rainfall in California was way below normal for the sixth year in a row and the reservoirs were drying up. Suggested solutions for the water shortage have included water rationing; building desalination, or desalting, plants to produce fresh water from seawater; and even importing water from Canada! Even if the farmers find some way to obtain more water for irrigation, problems remain.

The erosion of valuable topsoil as irrigation water runs off the land is one of those problems. Billions of tons of topsoil are washed from farmlands in the United States every year. Salts and toxic minerals from the water remain in the soil. These chemicals are harmful to plant growth. Eventually the amounts of salts and minerals increase to a level that kills plants. The soil becomes worthless for farming.

Water that runs off farmlands contains pesticides and fungicides as well as salts and minerals. This water then pollutes rivers, streams, lakes, marshes, and estuaries. Conservation groups warn that wildlife areas are frequently affected by this polluted runoff.

▲ **The parched land in this reservoir in northern California is evidence of the damage that can be caused by a drought.**

In some wildlife areas, plants and animals have died because of poisonous agricultural chemicals contained in the water that runs off farmlands.

In many parts of the West and Southwest, farmers face a double-headed problem. The need for fresh water increases steadily, even as the supply of fresh water is being depleted daily. Yet without irrigation, farmers cannot use their land to grow crops.

Farmers, ranchers, and developers say that a solution to this problem involves the development of new irrigation projects. They need a larger supply of water or their land will again dry up and become a desert.

Some environmentalists say that desert areas should no longer be irrigated—even if that means losing farms and ranches. Without irrigation, they argue, the land will return to its natural desert state. Plants and animals that are adapted to living in the deserts will return.

A number of state and federal water experts say that such a drastic solution is not necessary. They point out that the amount of water used to irrigate farmland could be cut drastically. Various techniques for saving water have been suggested. One irrigation system would deliver water directly to the plant roots. This method would cut down on the amount of water lost through evaporation and by runoff. Another suggestion involves covering crops with plastic to prevent evaporation. Controlling sprinklers by computers is another possible method of water conservation—water would be supplied only when it is needed. However, all these methods are expensive and would greatly increase the cost of irrigation water.

Many choices will have to be made about the future of desert irrigation. Should it continue? If so, under what conditions? Obviously, making the desert bloom involves thorny issues.

THE LONGEST WINTER

The team of Survivors shivered as the bouncing raft made its way across the river. An icy July wind whipped up the water all around the raft. The shore to which the Survivors were heading was covered with a thick coat of new snow. Although it was noon, the day was strangely dark.

Jonah, the team leader, peered into the eerie darkness ahead. Members of the team lowered long poles into the water and pushed the raft toward the shore.

"There," Jonah said, pointing to a good landing spot. "Head for that flat beach."

Within seconds, the team members were ashore. Soon after, they built a fire and cooked a meal of canned foods. And then they settled down for Jonah's daily afternoon "story" about the past.

"Well, my friends," Jonah began, "this day makes me think of a frightening story that began long, long ago. If I remember correctly, the time was the winter of 1983. The

football season was in full swing. The holidays were approaching. Everything was kind of normal and happy. Of course, people were worried about such things as taxes and war. But that wasn't too unusual for those days. Then something happened that scared a lot of people. A group of some of the finest scientists in the United States issued a report.

"The scientists—there were 20 of them—had been trying to figure out what might happen if there were a nuclear war. And the report described what they had discovered."

Jonah paused and looked up at the dark sky as if searching for the July sun. But all he could see was an unbroken gray cloud that seemed to stretch forever. Jonah sighed and went on.

"To get an idea of what the scientists concluded, you must first know some basic facts about this planet of ours. Our atmosphere is very special. For one thing, as one of the scientists said, 'our atmosphere normally acts as a window for sunlight but as a blanket for heat.'"

"What does that mean?" one of the Survivors asked as she tugged a blanket tightly around her shoulders.

"Normally, it means that the Earth stays warm. You see, the light from the sun is a form of energy. When this energy hits the Earth, a new kind of energy is produced. It is called heat. Our atmosphere normally traps a lot of this heat. The heat keeps us warm. It keeps our oceans and lakes from freezing. And it makes plant and animal life possible on the Earth. Now what do you think would happen if the 'window for sunlight' had a shade pulled down over it?"

A young Survivor, busily rubbing his hands together to keep them warm, answered, "The light from the sun would not reach the Earth. So heat energy would not be produced. And the Earth would cool down."

"Right," said Jonah. "And that's exactly what the scientists said could happen if there were a nuclear war. The scientists figured out that such a war would set millions of fires. Tons of black, sooty smoke would rise into the atmosphere. This smoke, they said, would have a mass of more than 100 million tons!

"The particles of soot would rise high into the atmosphere, the scientists calculated. As high, maybe, as 20 kilometers. There the

winds of the atmosphere would spread the soot over the entire Earth.

"Now there's something very special about these two things—the soot and the height to which it would rise. Experiments showed that small particles of black soot absorb sunlight better than other kinds of particles do. But that's not all. Researchers also found that the higher in the air particles are found, the longer they are going to stay there."

"That means that the soot closes the 'windows for sunlight.' And if the soot is high in the air, the 'window' stays closed for a long time," said one of the Survivors.

"Sadly, that's true," Jonah replied in a whisper. "Using the best computers of the time, the scientists figured out that about 95 percent of the sun's light would not reach the ground. In their report, the scientists said that in some places the brightness at noon 'could be as low as that of a moonlit night.'"

"Just like it is now," another Survivor said.

Jonah sighed. Telling this story wasn't easy, he thought to himself.

"So what would happen then?" asked another Survivor.

"Again, the computer came up with some answers. And they were chilling. Temperatures over land areas such as North America, Europe, and Asia would suddenly drop about 40°C! That means that the average temperature would be about –25°C."

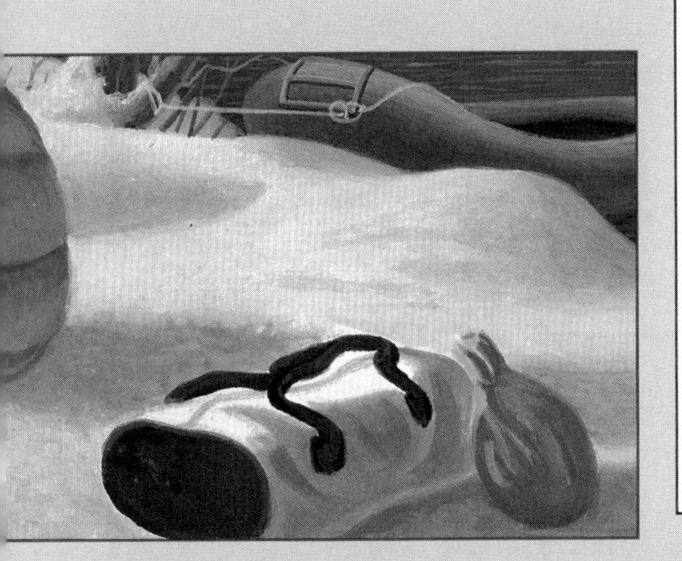

"Why that's way below the freezing point of water!" exclaimed a Survivor.

"I'm afraid so," said Jonah. "And these freezing temperatures would last for months, said the scientists. The scientists called it a 'nuclear winter.' And the winter would cover the whole world."

"But that would kill off all sorts of plants, along with the animals that live on plants," the Survivor continued. "Why, there would be no trees, no grass, no cows and sheep that feed on grass, no corn or wheat, no food for us."

"How about fishes in lakes and streams?" another Survivor asked.

"Frozen solid," said Jonah. "Life in the water would die out."

"But wouldn't living things survive in the oceans and on the coasts of oceans? After all, ocean water holds a lot of heat for a long time. It might keep the land nearby warm enough to grow crops," suggested a Survivor.

"The scientists thought of that too," replied Jonah. "But the very cold air over the land meeting the warm air over the sea would cause terrible storms. Farming in areas struck by constant hurricanes would be impossible."

Jonah stopped speaking. Again he looked up at the sky, searching for a ray of sunshine. But there was none to be seen. The Survivors sat quietly. Finally, one of them spoke.

"If that had happened, Jonah, the whole world would have looked and felt like this day. Everywhere it would be winter in July—not just here in Argentina and other southern countries. And we would be real survivors instead of members of an outdoor club."

"That's right," said Jonah, who now smiled. "But leaders of our country and of the other countries of the world saw to it that there was no nuclear war. So we still have our trees and our grass, our forests, our animals... and our wilderness to explore. So let's pack up our gear and be on our way. We've got a long hike ahead of us."

For Further Reading

If you have been intrigued by the concepts examined in this textbook, you may also be interested in the ways fellow thinkers—novelists, poets, essayists, as well as scientists—have imaginatively explored the same ideas.

Chapter 1: What Is Weather?

Aaron, Chester. *An American Ghost.* New York: Harcourt, Brace, Jovanovich.

Babbitt, Natalie. *The Eyes of the Amaryllis.* New York: Farrar, Straus and Giroux.

Mayo, Gretchen. *Earthmaker's Tales: North American Indian Stories About Earth Happenings.* New York: Walker

Southall, Ivan. *Hill's End.* New York: Macmillan

Chapter 2: What Is Climate?

Clark, Mavis Thorpe. *Wildfire.* New York: Macmillan.

O'Dell, Scott. *Island of the Blue Dolphins.* Boston: Houghton Mifflin Co.

Skurzynski, Gloria. *Trapped in the Slickrock Canyon.* New York: Lothrop, Lee & Shepard Books.

Strieber, Whitley. *Wolf of Shadows.* New York: Alfred A. Knopf.

Chapter 3: Climate in the United States

Dyer, T.A. *A Way of His Own.* Boston: Houghton Mifflin Co.

George, Jean Craighead. *Julie of the Wolves.* New York: Harper & Row.

Turner, Ann. *Grasshopper Summer.* New York: Macmillan.

Activity Bank

Welcome to the Activity Bank! This is an exciting and enjoyable part of your science textbook. By using the Activity Bank you will have the chance to make a variety of interesting and different observations about science. The best thing about the Activity Bank is that you and your classmates will become the detectives, and as with any investigation you will have to sort through information to find the truth. There will be many twists and turns along the way, some surprises and disappointments too. So always remember to keep an open mind, ask lots of questions, and have fun learning about science.

HOW CAN YOU PREVENT FOOD FROM SPOILING?

The growth of bacteria, which can cause packaged foods to spoil, can be prevented by the use of ultraviolet light. What are some other ways in which food manufacturers can slow down or stop the growth of bacteria in their food products? To stop the growth of bacteria, it is necessary to take away one or more of the conditions bacteria need in order to grow. In this activity you will investigate some of these conditions and find out how they are changed to prevent bacterial growth. All you will need are packages and labels from different food products.

1. Carefully examine each food product. Identify each food product and describe the way in which it is packaged. For example, is the food in a glass jar, an aluminum can, a cardboard box, or some other type of package? Is the food frozen? Include as much information as you can about each food.
2. Read the package label for each food product. According to the label, what has been done to prevent bacteria from growing in each food? For example, has salt been added to the food? Bacteria cannot grow if too much salt is present. So adding salt is one way to prevent food from spoiling.
3. Share your results with the entire class. Based on your observations of the food packages and labels, what are some conditions necessary for bacterial growth? How are these conditions changed to slow down or prevent bacterial growth in packaged food?

Think for Yourself

If you have ever canned fruits or vegetables at home, you know that the food must first be heated and then sealed in airtight jars. How does this process prevent bacterial growth?

BUILD YOUR OWN ANEMOMETER

Meteorologists use an anemometer to measure wind speed. An anemometer usually has three or four cups at the ends of horizontal arms that are attached to a vertical shaft. When the wind blows into the cups, the arms turn and a meter records the wind speed in miles per hour. In this activity you will build your own simple anemometer.

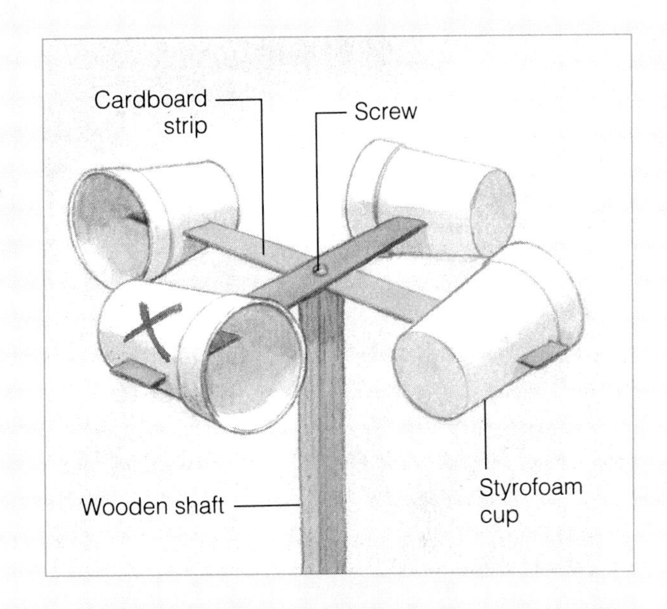

Materials

2 strips of cardboard, 10 cm x 30 cm
scissors
metric ruler
4 Styrofoam cups
marking pen
stapler
2 washers
screw
screwdriver
wooden shaft, 30 cm long
clock or watch with second hand

Procedure

1. With a marking pen, draw a large X on one of the Styrofoam cups.
2. Carefully cut vertical slits in the front and back of each cup. Make the slits wide enough so that you can push a cardboard strip through them.
3. Make an X by placing one cardboard strip on top of the other. Staple the strips together.
4. Attach the four cups to the four arms of the X by sliding the cardboard strips through the slits in the cups. The openings of the cups should face counterclockwise.
5. To mount the X on the wooden shaft, make a hole in the center of the X. Place a washer on the screw and push the screw through the hole in the X. Place the second washer on the screw below the X and screw the X to the top of the shaft. Be sure that your anemometer can turn freely.
6. Position your anemometer in a spot where the wind can hit it from all directions. To measure the wind speed, count the number of times the anemometer turns in 30 seconds and divide by 5. The result is the wind speed in miles per hour. (*Hint:* Use the cup marked with an X as a guide.)

Do It Yourself

As a class project, you might want to set up your own weather observation station. What other instruments, in addition to an anemometer, could you add to your weather station to help you make weather observations and forecasts?

WHAT CAUSES LIGHTNING?

As Benjamin Franklin discovered, lightning is a form of electricity. Lightning is caused when electric charges build up in storm clouds, resulting in a discharge, or flash of lightning. There are two kinds of electric charges: positive and negative. In this activity you will investigate what happens when electrically charged objects are brought together.

1. To begin your investigation, obtain two balloons, two pieces of string, a glass rod, and a silk scarf.

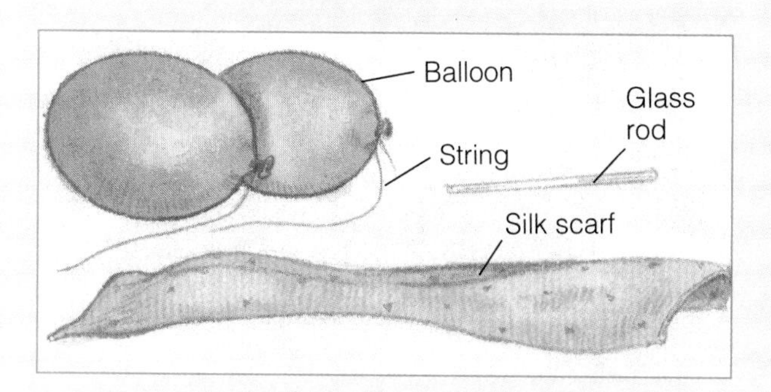

2. Blow up the balloons and tie a piece of string to each balloon.
3. Rub each balloon with the silk scarf. Hold the balloons by the string and bring them near one another. What happens to the balloons? Rubbing the balloons with silk gave each balloon a positive electric charge. Based on your observations, what happens when two objects with like charges are brought together?
4. Rub one balloon again with the silk scarf. Hold the scarf near the balloon. What happens? Rubbing the balloon with silk gave the balloon a positive charge and the silk a negative charge. Based on your observations, what happens when two objects with unlike charges are brought together?

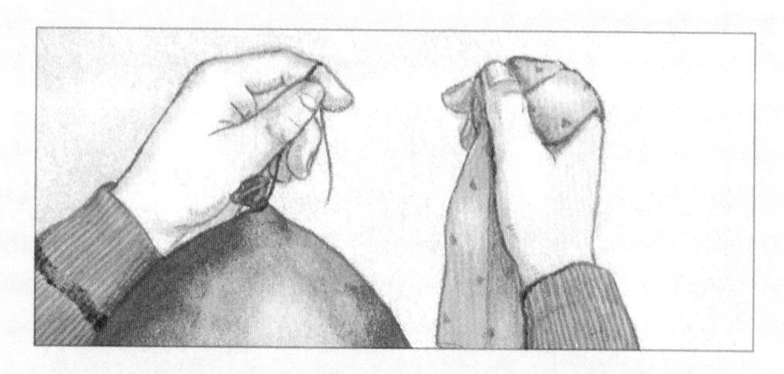

5. Rub the glass rod with the silk scarf. What kind of electric charge does the glass rod have after being rubbed with silk?
6. Turn on a stream of water from a faucet. The water is neutral—that is, it has no electric charge. What do you predict will happen if you bring the charged glass rod near the stream of water? Try it and find out. Was your prediction correct? Based on your observations, what happens when a charged object is brought near an uncharged (neutral) object?

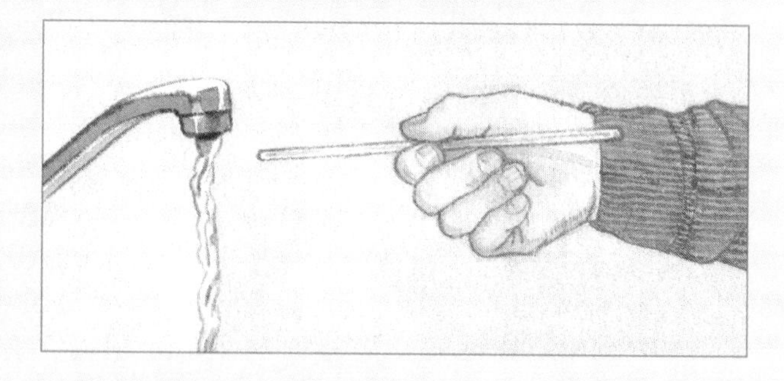

Think for Yourself

Has this ever happened to you? You walk across a room with a thick wool carpet on the floor and reach out to touch a metal doorknob. As you touch the doorknob, you feel a tingle in your fingertips. Based on what you now know about electric charges, what do you think causes the tingle you feel?

WHAT ARE DENSITY CURRENTS?

An ocean current is a river of water in the ocean. Ocean currents caused by wind patterns are called surface currents. Surface currents are either cold-water currents or warm-water currents. Other ocean currents, called deep currents, are caused mainly by differences in the density of ocean water. In this activity you will investigate how adding salt to water affects the density of water.

Materials

200-mL beaker
clear plastic container
salt
plastic spoon
food coloring, blue and red
sheet of paper

Procedure

1. Fill a 200-mL beaker with water. Add some blue food coloring to the water. Add a spoonful of salt and stir to dissolve.

2. Fill the plastic container with water. Add some red food coloring and stir.
3. Tear a sheet of paper into small pieces. Sprinkle several pieces of paper onto the surface of the water in the container.

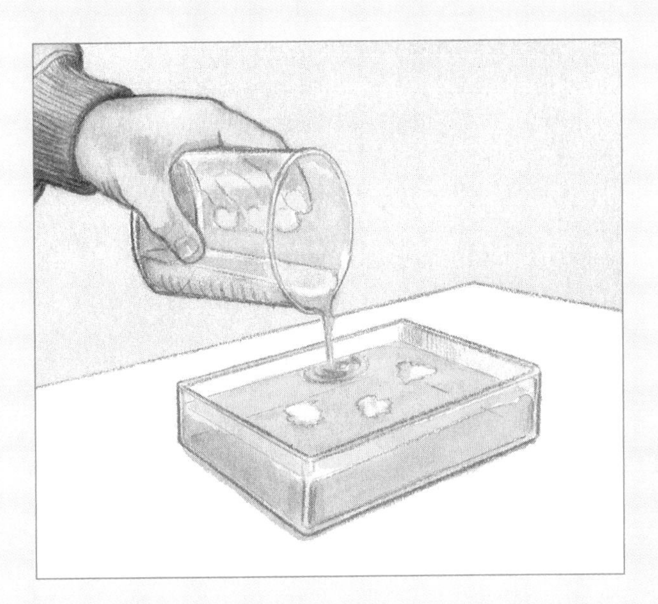

4. Gently pour the salt water down one side of the container. What happens? Do the pieces of paper move?
5. Leave the container undisturbed and continue to make observations every 5 minutes for the next 20 to 30 minutes. What happened when you poured salt water into fresh water? What effect does adding salt have on the density of water? What caused the paper to move?

Going Further

Repeat steps 1 to 3. To a second beaker, add water, a spoonful of salt, a few ice cubes, and some yellow food coloring. Pour the blue salt water into the container as you did in step 4. Then add the yellow salt water in the same way. How many layers do you see in the container?

WHAT IS YOUR LATITUDE?

Early sailors used an instrument called an astrolabe to help them navigate across open waters on long ocean voyages. An astrolabe measures the altitude, or angular distance, of Polaris, the North Star, above the horizon. This angular distance, in degrees, is equal to the latitude at the observer's location. In this activity you will make a simple astrolabe and use it to determine the altitude of Polaris at your location. This will give you a good approximation of your latitude.

Materials

protractor
piece of cardboard
small weight
drinking straw
tape
string, 30 cm
thumbtack

Procedure

1. Tape a protractor to a small piece of cardboard.
2. Tape a drinking straw to the straight edge of the protractor.
3. Use a thumbtack to attach a piece of string to the center hole of the protractor. Tie a small weight to the free end of the string.
4. On a clear night, go outside and locate Polaris. (*Hint*: You might want to use a star chart to help you find Polaris.) Sight Polaris through the straw as shown.
5. When you have sighted Polaris, press the string against the protractor and read the number of degrees on the protractor. The number of degrees is equal to the altitude of Polaris and also to your approximate latitude. What is your latitude?
6. Compare your observed latitude with the actual latitude listed in an atlas for your location. How accurate was your astrolabe measurement? What might have caused any difference between your observed latitude and the actual latitude? How could you improve the accuracy of your measurement?

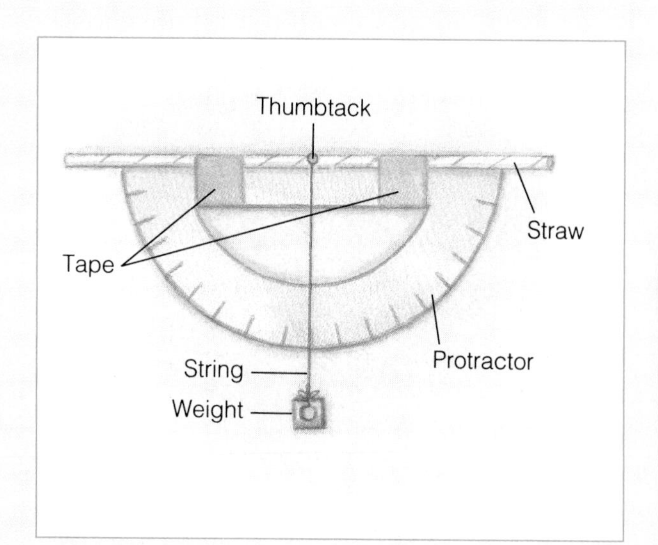

EARTH'S ELLIPTICAL ORBIT

The shape of the Earth's orbit, like the orbits of all the other planets in the solar system, is an ellipse. An ellipse is a geometrical figure drawn around two fixed points, or foci (singular: focus). For an elliptical planetary orbit, one focus is the sun. The other focus is a point in space. In this activity you will explore how the shape of the Earth's orbit affects its distance from the sun. You will need a piece of cardboard, two thumbtacks, a piece of string, and a pencil.

1. Stick two thumbtacks into a piece of cardboard. Place the thumbtacks about 10 cm apart. The thumbtacks will represent the foci of your ellipse. Label one focus "Sun."
2. Use a piece of string 30 cm long. Tie the ends of the string together and wind the string around the thumbtacks as shown.
3. With a sharp pencil, trace an ellipse on the cardboard. Be sure to keep the string taut as you draw the ellipse.
4. Repeat steps 1 to 3, but this time place the thumbtacks 5 cm apart. Compare the shapes of the two ellipses you have drawn. How does the distance between the foci affect the shape of an ellipse? Which elliptical orbit would bring the Earth closer to the sun? How would this affect the amount of radiant energy the Earth received? How would changing the amount of radiant energy affect the Earth's climate?

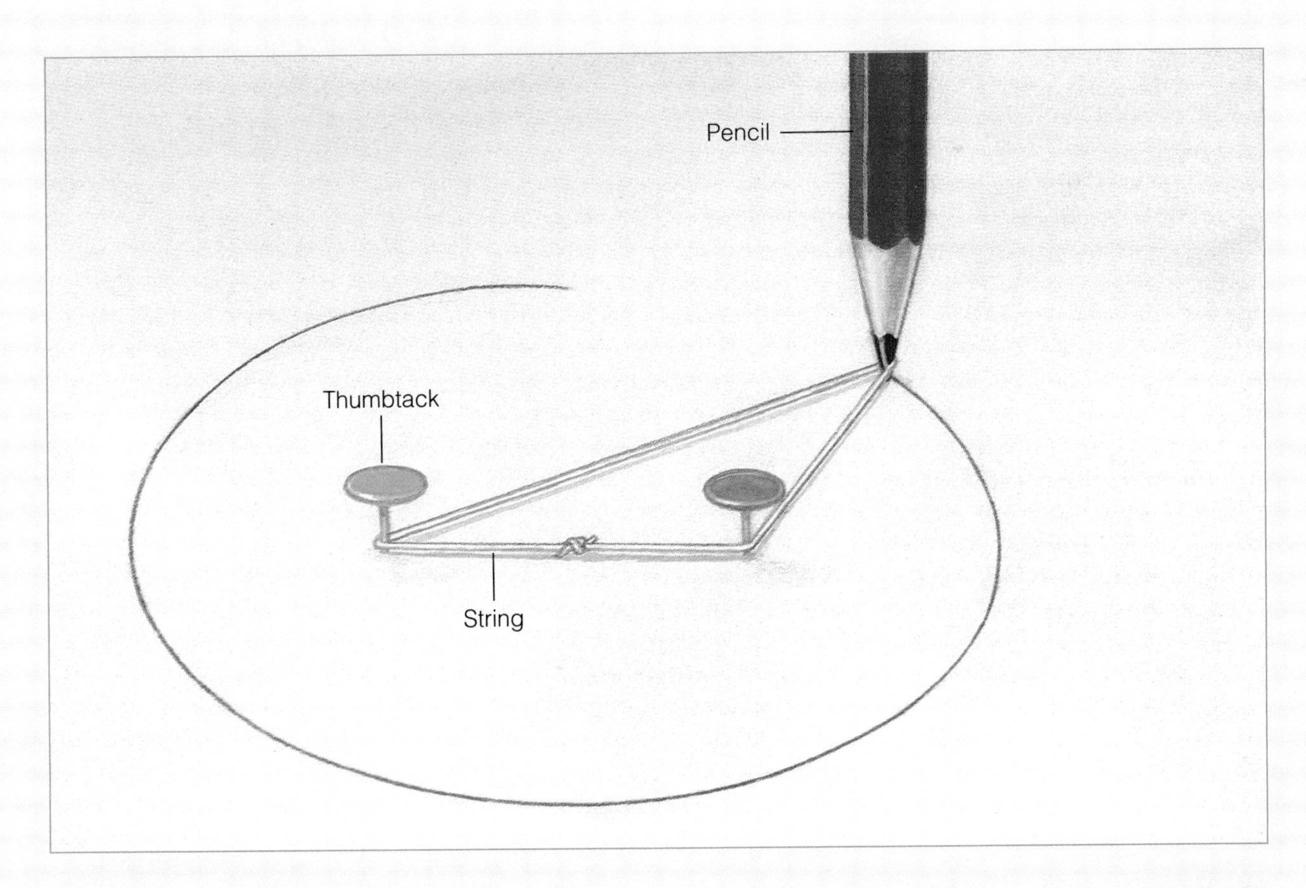

SOIL PERMEABILITY

The types of plants that can grow in different biomes depend to some extent on the type of soil present in those biomes. For example, because of the permanently frozen layer of soil called permafrost, plant growth in a tundra biome is limited to low-growing shrubs, mosses, and grasses. There are many different types of soil. Some soils soak up water better than others. The ability of soil to soak up water is called permeability. In this activity you will compare the permeability of different soils.

Materials

3 aluminum cans
can opener
100-mL graduated cylinder
clock or watch with second hand

Procedure

1. With your teacher's permission, choose three different locations on the school grounds where you can test the soil.
2. Remove the tops and bottoms from three empty aluminum soft drink cans.
3. In each location, press a can into the soil up to about 3 cm from the top of the can. **CAUTION:** *Use your foot to press the cans into the soil. Do not use your hands.*
4. Fill a 100-mL graduated cylinder with water. Pour the water into one of the cans.

5. Measure how much time it takes for the water in the can to soak into the soil. Record your observations.
6. Repeat steps 4 and 5 for the other two cans. Compare your results. Which location had the most permeable soil? The least? How can you tell? How might the permeability of the soil in a particular biome affect plant growth in that biome?

Appendix A

THE METRIC SYSTEM

The metric system of measurement is used by scientists throughout the world. It is based on units of ten. Each unit is ten times larger or ten times smaller than the next unit. The most commonly used units of the metric system are given below. After you have finished reading about the metric system, try to put it to use. How tall are you in metrics? What is your mass? What is your normal body temperature in degrees Celsius?

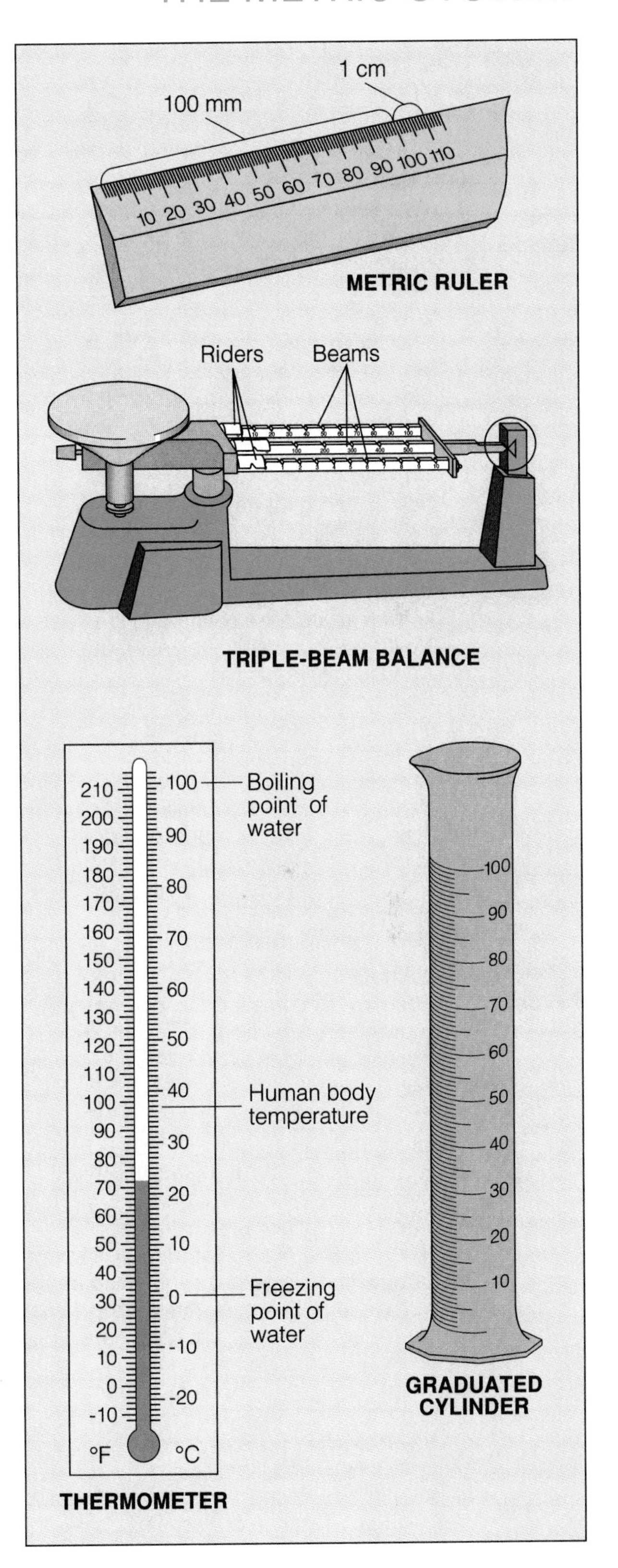

Commonly Used Metric Units

Length The distance from one point to another

meter (m) A meter is slightly longer than a yard.
1 meter = 1000 millimeters (mm)
1 meter = 100 centimeters (cm)
1000 meters = 1 kilometer (km)

Volume The amount of space an object takes up

liter (L) A liter is slightly more than a quart.
1 liter = 1000 milliliters (mL)

Mass The amount of matter in an object

gram (g) A gram has a mass equal to about one paper clip.
1000 grams = 1 kilogram (kg)

Temperature The measure of hotness or coldness

degrees Celsius (°C) 0°C = freezing point of water
100°C = boiling point of water

Metric–English Equivalents

2.54 centimeters (cm) = 1 inch (in.)
1 meter (m) = 39.37 inches (in.)
1 kilometer (km) = 0.62 miles (mi)
1 liter (L) = 1.06 quarts (qt)
250 milliliters (mL) = 1 cup (c)
1 kilogram (kg) = 2.2 pounds (lb)
28.3 grams (g) = 1 ounce (oz)
$°C = 5/9 \times (°F - 32)$

Appendix B

LABORATORY SAFETY
Rules and Symbols

Glassware Safety

1. Whenever you see this symbol, you will know that you are working with glassware that can easily be broken. Take particular care to handle such glassware safely. And never use broken or chipped glassware.
2. Never heat glassware that is not thoroughly dry. Never pick up any glassware unless you are sure it is not hot. If it is hot, use heat-resistant gloves.
3. Always clean glassware thoroughly before putting it away.

Fire Safety

1. Whenever you see this symbol, you will know that you are working with fire. Never use any source of fire without wearing safety goggles.
2. Never heat anything—particularly chemicals—unless instructed to do so.
3. Never heat anything in a closed container.
4. Never reach across a flame.
5. Always use a clamp, tongs, or heat-resistant gloves to handle hot objects.
6. Always maintain a clean work area, particularly when using a flame.

Heat Safety

Whenever you see this symbol, you will know that you should put on heat-resistant gloves to avoid burning your hands.

Chemical Safety

1. Whenever you see this symbol, you will know that you are working with chemicals that could be hazardous.
2. Never smell any chemical directly from its container. Always use your hand to waft some of the odors from the top of the container toward your nose—and only when instructed to do so.
3. Never mix chemicals unless instructed to do so.
4. Never touch or taste any chemical unless instructed to do so.
5. Keep all lids closed when chemicals are not in use. Dispose of all chemicals as instructed by your teacher.
6. Immediately rinse with water any chemicals, particularly acids, that get on your skin and clothes. Then notify your teacher.

Eye and Face Safety

1. Whenever you see this symbol, you will know that you are performing an experiment in which you must take precautions to protect your eyes and face by wearing safety goggles.
2. When you are heating a test tube or bottle, always point it away from you and others. Chemicals can splash or boil out of a heated test tube.

Sharp Instrument Safety

1. Whenever you see this symbol, you will know that you are working with a sharp instrument.
2. Always use single-edged razors; double-edged razors are too dangerous.
3. Handle any sharp instrument with extreme care. Never cut any material toward you; always cut away from you.
4. Immediately notify your teacher if your skin is cut.

Electrical Safety

1. Whenever you see this symbol, you will know that you are using electricity in the laboratory.
2. Never use long extension cords to plug in any electrical device. Do not plug too many appliances into one socket or you may overload the socket and cause a fire.
3. Never touch an electrical appliance or outlet with wet hands.

Animal Safety

1. Whenever you see this symbol, you will know that you are working with live animals.
2. Do not cause pain, discomfort, or injury to an animal.
3. Follow your teacher's directions when handling animals. Wash your hands thoroughly after handling animals or their cages.

Appendix C

SCIENCE SAFETY RULES

One of the first things a scientist learns is that working in the laboratory can be an exciting experience. But the laboratory can also be quite dangerous if proper safety rules are not followed at all times. To prepare yourself for a safe year in the laboratory, read over the following safety rules. Then read them a second time. Make sure you understand each rule. If you do not, ask your teacher to explain any rules you are unsure of.

Dress Code

1. Many materials in the laboratory can cause eye injury. To protect yourself from possible injury, wear safety goggles whenever you are working with chemicals, burners, or any substance that might get into your eyes. Never wear contact lenses in the laboratory.

2. Wear a laboratory apron or coat whenever you are working with chemicals or heated substances.

3. Tie back long hair to keep it away from any chemicals, burners and candles, or other laboratory equipment.

4. Remove or tie back any article of clothing or jewelry that can hang down and touch chemicals and flames.

General Safety Rules

5. Read all directions for an experiment several times. Follow the directions exactly as they are written. If you are in doubt about any part of the experiment, ask your teacher for assistance.

6. Never perform activities that are not authorized by your teacher. Obtain permission before "experimenting" on your own.

7. Never handle any equipment unless you have specific permission.

8. Take extreme care not to spill any material in the laboratory. If a spill occurs, immediately ask your teacher about the proper cleanup procedure. Never simply pour chemicals or other substances into the sink or trash container.

9. Never eat in the laboratory.

10. Wash your hands before and after each experiment.

First Aid

11. Immediately report all accidents, no matter how minor, to your teacher.

12. Learn what to do in case of specific accidents, such as getting acid in your eyes or on your skin. (Rinse acids from your body with lots of water.)

13. Become aware of the location of the first-aid kit. But your teacher should administer any required first aid due to injury. Or your teacher may send you to the school nurse or call a physician.

14. Know where and how to report an accident or fire. Find out the location of the fire extinguisher, phone, and fire alarm. Keep a list of important phone numbers—such as the fire department and the school nurse—near the phone. Immediately report any fires to your teacher.

Heating and Fire Safety

15. Again, never use a heat source, such as a candle or burner, without wearing safety goggles.

16. Never heat a chemical you are not instructed to heat. A chemical that is harmless when cool may be dangerous when heated.

17. Maintain a clean work area and keep all materials away from flames.

18. Never reach across a flame.

19. Make sure you know how to light a Bunsen burner. (Your teacher will demonstrate the proper procedure for lighting a burner.) If the flame leaps out of a burner toward you, immediately turn off the gas. Do not touch the burner. It may be hot. And never leave a lighted burner unattended!

20. When heating a test tube or bottle, always point it away from you and others. Chemicals can splash or boil out of a heated test tube.

21. Never heat a liquid in a closed container. The expanding gases produced may blow the container apart, injuring you or others.

22. Before picking up a container that has been heated, first hold the back of your hand near it. If you can feel the heat on the back of your hand, the container may be too hot to handle. Use a clamp or tongs when handling hot containers.

Using Chemicals Safely

23. Never mix chemicals for the "fun of it." You might produce a dangerous, possibly explosive substance.

24. Never touch, taste, or smell a chemical unless you are instructed by your teacher to do so. Many chemicals are poisonous. If you are instructed to note the fumes in an experiment, gently wave your hand over the opening of a container and direct the fumes toward your nose. Do not inhale the fumes directly from the container.

25. Use only those chemicals needed in the activity. Keep all lids closed when a chemical is not being used. Notify your teacher whenever chemicals are spilled.

26. Dispose of all chemicals as instructed by your teacher. To avoid contamination, never return chemicals to their original containers.

27. Be extra careful when working with acids or bases. Pour such chemicals over the sink, not over your workbench.

28. When diluting an acid, pour the acid into water. Never pour water into an acid.

29. Immediately rinse with water any acids that get on your skin or clothing. Then notify your teacher of any acid spill.

Using Glassware Safely

30. Never force glass tubing into a rubber stopper. A turning motion and lubricant will be helpful when inserting glass tubing into rubber stoppers or rubber tubing. Your teacher will demonstrate the proper way to insert glass tubing.

31. Never heat glassware that is not thoroughly dry. Use a wire screen to protect glassware from any flame.

32. Keep in mind that hot glassware will not appear hot. Never pick up glassware without first checking to see if it is hot. See #22.

33. If you are instructed to cut glass tubing, fire-polish the ends immediately to remove sharp edges.

34. Never use broken or chipped glassware. If glassware breaks, notify your teacher and dispose of the glassware in the proper trash container.

35. Never eat or drink from laboratory glassware. Thoroughly clean glassware before putting it away.

Using Sharp Instruments

36. Handle scalpels or razor blades with extreme care. Never cut material toward you; cut away from you.

37. Immediately notify your teacher if you cut your skin when working in the laboratory.

Animal Safety

38. No experiments that will cause pain, discomfort, or harm to mammals, birds, reptiles, fishes, and amphibians should be done in the classroom or at home.

39. Animals should be handled only if necessary. If an animal is excited or frightened, pregnant, feeding, or with its young, special handling is required.

40. Your teacher will instruct you as to how to handle each animal species that may be brought into the classroom.

41. Clean your hands thoroughly after handling animals or the cage containing animals.

End-of-Experiment Rules

42. After an experiment has been completed, clean up your work area and return all equipment to its proper place.

43. Wash your hands after every experiment.

44. Turn off all burners before leaving the laboratory. Check that the gas line leading to the burner is off as well.

Appendix D

WEATHER MAP SYMBOLS

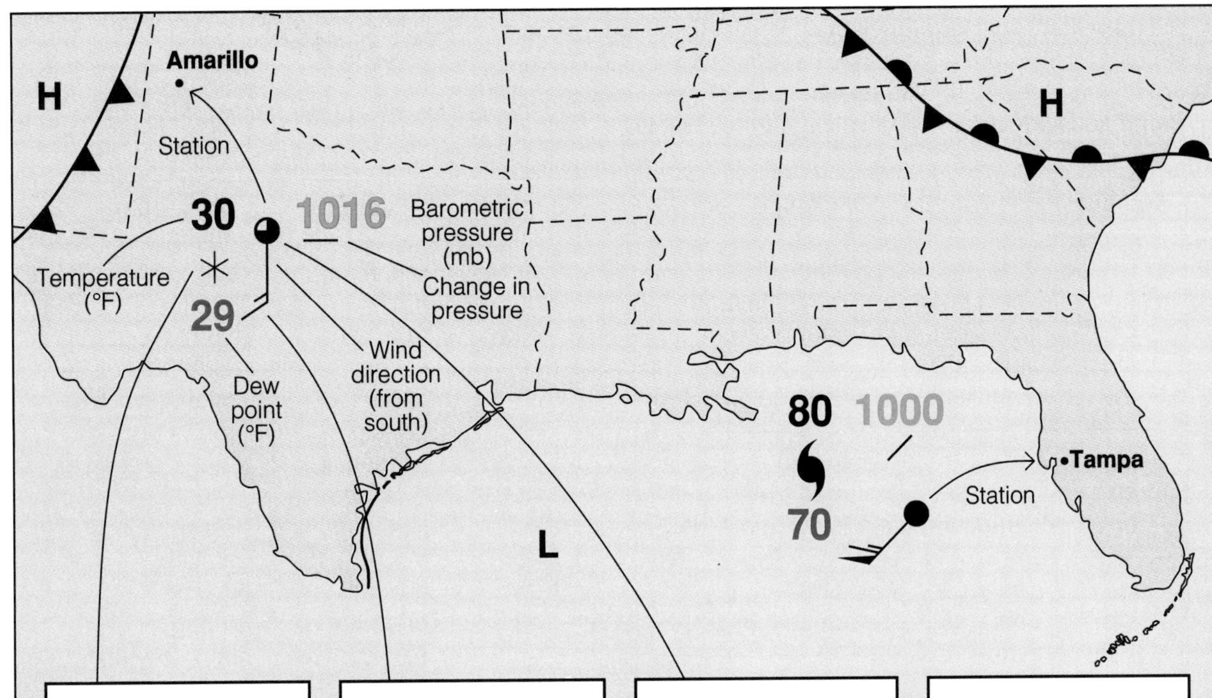

Weather	Symbol
Drizzle	
Fog	
Hail	
Haze	
Rain	
Shower	
Sleet	
Smoke	
Snow	
Thunderstorm	
Hurricane	

Wind Speed (mph)	Symbol
1–4	
5–8	
9–14	
15–20	
21–25	
26–31	
32–37	
38–43	
44–49	
50–54	
55–60	
61–66	
67–71	
72–77	

Cloud Cover (%)	Symbol
0	
10	
20–30	
40	
50	
60	
70–80	
90	
100	

Fronts and Pressure Systems	Symbol
Cold front	
Warm front	
Stationary front	
Occluded front	
High pressure	H
Low pressure	L
Rising	
Steady	
Falling	

Glossary

Pronunciation Key

When difficult names or terms first appear in the text, they are respelled to aid pronunciation. A syllable in SMALL CAPITAL LETTERS receives the most stress. The key below lists the letters used for respelling. It includes examples of words using each sound and shows how the words would be respelled.

Symbol	Example	Respelling
a	hat	(hat)
ay	pay, late	(pay), (layt)
ah	star, hot	(stahr), (haht)
ai	air, dare	(air), (dair)
aw	law, all	(law), (awl)
eh	met	(meht)
ee	bee, eat	(bee), (eet)
er	learn, sir, fur	(lern), (ser), (fer)
ih	fit	(fiht)
igh	mile, sigh	(mighl), (sigh)
oh	no	(noh)
oi	soil, boy	(soil), (boi)
oo	root, tule	(root), (rool)
or	born, door	(born), (dor)
ow	plow, out	(plow), (owt)

Symbol	Example	Respelling
u	put, book	(put), (buk)
uh	fun	(fuhn)
yoo	few, use	(fyoo), (yooz)
ch	chill, reach	(chihl), (reech)
g	go, dig	(goh), (dihg)
j	jet, gently, bridge	(jeht), (JEHNT-lee), (brihj)
k	kite, cup	(kight), (kuhp)
ks	mix	(mihks)
kw	quick	(kwihk)
ng	bring	(brihng)
s	say, cent	(say), (sehnt)
sh	she, crash	(shee), (krash)
th	three	(three)
y	yet, onion	(yeht), (UHN-yuhn)
z	zip, always	(zihp), (AWL-wayz)
zh	treasure	(TREH-zher)

air mass: large body of air with uniform properties throughout

air pressure: measure of the force of air pressing down on the Earth's surface

anemometer (an-uh-MAHM-uh-ter): instrument used to measure wind speed

atmosphere (AT-muhs-feer): mixture of gases that surrounds the Earth

barometer (buh-RAHM-uh-ter): instrument used to measure air pressure

biome (BIGH-ohm): division used to classify areas with similar climates, plants, and animals

canopy (KAN-uh-pee): top layer of tropical rain forest

climate: general conditions of temperature and precipitation for an area over a long period of time

conduction: direct transfer of heat energy from one substance to another

conifer: needle-leaved tree that produces its seeds in cones

continental climate: climate found in areas within a large landmass

convection: transfer of heat energy in a fluid (gas or liquid)

Coriolis effect: shift in wind direction caused by the rotation of the Earth on its axis

desert climate: climate found in the western interior of the United States; characterized by very low precipitation and high temperatures in summer and winter

evaporation: process by which radiant energy from the sun turns liquid water into a gas (water vapor)

front: boundary that forms when two air masses with different properties meet

greenhouse effect: process in which carbon dioxide and other gases in the atmosphere absorb infrared radiation from the sun, forming a "heat blanket" around the Earth

interglacial: time period between major glaciations (or ice ages)

isobar (IGH-soh-bar): line on a weather map that connects locations with the same air pressure

isotherm (IGH-so-therm): line on a weather map that connects locations with the same temperature

land breeze: flow of air from the land to the sea

leeward side: side of a mountain facing away from the wind

major glaciation: period in the Earth's history when large parts of the Earth's surface were covered with sheets of ice; ice age

marine climate: climate found in areas near an ocean or other large body of water

marine west coast climate: climate found along the northwestern coast of the United States; characterized by heavy precipitation, mild winters, and cool summers

Mediterranean climate: climate found in the coastal area of California; characterized by heavy precipitation in winter but dry summers, with summer temperatures only slightly higher than winter temperatures

microclimate: small, localized climate

moist continental climate: climate found from the northern Midwest to the Atlantic coast of the United States; characterized by a moderate amount of precipitation all year, with very cold winters and hot summers

moist subtropical climate: climate found in the southeastern part of the United States; characterized by more precipitation in summer than in winter, with hot summers and mild winters

permafrost: permanently frozen layer of soil on a tundra

polar zone: climate zone extending from the pole (90°) to about 60° latitude in each hemisphere

precipitation (pree-sihp-uh-TAY-shuhn): water that falls from the atmosphere to the Earth as rain, sleet, snow, or hail

prevailing wind: wind that blows more often from one direction than from any other direction

psychrometer (sigh-KRAHM-uh-ter): instrument used to measure relative humidity

radiation: transfer of heat energy through empty space

rain gauge: instrument used to measure rainfall

relative humidity: percentage of moisture the air holds relative to the amount it could hold at a particular temperature

sea breeze: flow of air from the sea to the land

steppe climate: climate found in the western interior of the United States; similar to a desert climate but with slightly more precipitation

taiga (TIGH-guh): another name for a coniferous forest biome; Russian word that means swamp forest

temperate zone: climate zone located between 60° and 30° latitude in each hemisphere

thermometer: instrument used to measure temperature

tropical zone: climate zone located between 30° latitude and the equator (0°) in each hemisphere

wind: movement of air from an area of high pressure to an area of lower pressure

windward side: side of a mountain facing toward the wind

Index

Credits

Cover Background: Ken Karp
Photo Research: Omni-Photo Communications, Inc.
Contributing Artists: Michael Adams/Phil Veloric, Art Representatives; Anni Matsick/Cornell and McCarthy Art Representatives; Ray Smith; Warren Budd Assoc. Ltd.; Don Martinetti; David Biedrzycki; Gerry Schrenk
Photographs: **4** left: D. Cavagnaro/DRK Photo; right: Tony Stone Worldwide/Chicago Ltd.; **5** top:NASA/Omni-Photo Communications, Inc.; bottom: Steven Wayne Rotsch/DPI; **6** top: Lefever/Grushow Grant Heilman Photography; center: Index Stock Photography, Inc.; bottom: Rex Joseph; **8** top: NASA; center: Nubar Alexanian/Woodfin Camp & Associates; bottom: Odyssey/Woodfin Camp & Associates; **9** left: S. Rotner/Omni-Photo Communications, Inc.; right: Mark Lewis/Tony Stone Worldwide/Chicago Ltd.; **10** and **11** Stephen J. Krasemann/DRK Photo; **12** left: Chris Hackett/Image Bank; right: Kenneth Garrett/Woodfin Camp & Associates; **16** top: John Cancalosi/DRK Photo; bottom left: NASA; bottom right: Ray Pfortner/Peter Arnold, Inc.; **17** left: DPI; center: Steve Vidler/Leo De Wys, Inc.; right: Tony Stone Worldwide/Chicago Ltd.; **19** James Balog/Tony Stone Worldwide/Chicago Ltd.; **21** Runk/Schoenberger/Grant Heilman Photography; **22** Tony Stone Worldwide/Chicago Ltd.; **26** top: NASA/Omni-Photo Communications, Inc.; bottom: Harriet Arnold/DPI; **27** top: NASA; bottom: Frank LaBua/Envision; **28** Gene Moore; **29** left: Keith Olson/Tony Stone Worldwide/Chicago Ltd.; right: E. Spiegelhalter/Woodfin Camp & Associates; **31** left: Tom Bean/Stock Market; center: Grant Heilman/Grant Heilman Photography; right: G. R. Roberts/Omni-Photo Communications, Inc.; **33** Beryl Bidwell/Tony Stone Worldwide/Chicago Ltd.; **34** top: John H. Gerard/DPI; center: Tony Stone Worldwide/Chicago Ltd.; bottom: Runk/Schoenberger/Grant Heilman Photography; **37** and **38** NOAA/NESDIS; **39** left: Sumner/Stock Market; right: Annie Griffiths/DRK Photo; **40** H. Wendler/Image Bank; **41** left: NASA; right: B.Wisser/Gamma-Liasison, Inc.; **42** top: A&J Verkaik/Stock Market; bottom: Charles Gupton/Stock Market; **43** top: Howard B. Bluestein/University of Oklahoma/Photo Researchers, Inc.; bottom: Bill Bachman/Science Source/Photo Researchers, Inc.; **45** Stephen J. Krasemann/DRK Photo; **47** M. Long/Envision; **48** left: Culver Pictures, Inc.; right: Jim Brandenburg/Woodfin Camp & Associates; **49** top: N. H. (Dan) Cheatham/DRK Photo; bottom: Anthony Howarth/Woodfin Camp & Associates; **56** top: The Stone Flower Studio/DPI; bottom: DPI; **57** J. Leidmann/Leo De Wys, Inc.; **58** NOAA/NESDIS/NCDC/SOSD U.S. Department of Commerce; **59** Tony Stone Worldwide/Chicago Ltd.; **61** top: Lee L. Waldman/Stock Market; bottom: Grace Davies/Envision; **63** Tony Stone Worldwide/Chicago Ltd.; **64** left: Stan Osolinski/Stock Market; right: Wayne Lynch/DRK Photo; **65** left: Envision; right: R. Hackett/Omni-Photo Communications, Inc.; **66** Hank Morgan/Rainbow; **68** left: Peter Hendrie/Image Bank; right: Steve Leonard/Tony Stone Worldwide/Chicago Ltd.; **73** top: Walter Geiersperger/Tony Stone Worldwide/Chicago Ltd.; bottom: Stephanie Maze/Woodfin Camp & Associates; **74** top left: Roger Job/Gamma-Liaison, Inc.; bottom left: R. Dobbs/Gamma-Liaison, Inc.; **82** and **83** Johnny Johnson/DRK Photo; **84** top: Joseph Brignolo/Image Bank; bottom left: David Muench Photography Inc.; bottom right: Garry D. McMichael/Photo Researchers, Inc.; **86** left: Thomas J. Styczynski/Tony Stone Worldwide/Chicago Ltd.; right: David W. Hamilton/Image Bank; **87** left: David Muench Photography Inc.; right: Zigy Kalunzy/Tony Stone Worldwide/Chicago Ltd.; **88** left and right: David Muench Photography Inc.; **90** left: M. P. Kahl/DRK Photo; right: Bob Thomason/Tony Stone Worldwide/Chicago Ltd.; **91** left: Don Klumpp/Image Bank; center: Brian Parker/Tom Stack & Associates; right: Larry Lipsky/Tom Stack & Associates; **92** left: David Muench Photography Inc.; right: Shattil/Rozinski/Tom Stack & Associates; **95** Thomas Kitchin/Tom Stack & Associates; **96** top: Michael P. Gadomski/Animals Animals/Earth Scenes; center: John Shaw/Tom Stack & Associates; bottom left: Charles G. Summers, Jr./DPI; bottom center: Wayne Lynch/DRK Photo; bottom right: Tom Ulrich/Tony Stone Worldwide/Chicago Ltd.; **97** top: Chuck Place/Image Bank; bottom left: Tony Stone Worldwide/Chicago Ltd.; bottom center: Johnny Johnson/DRK Photo; bottom right: Phil Dotson/DPI; **98** top: Larry Ulrich/DRK Photo; bottom left: Wayne Lankinen/DRK Photo; bottom center: Breck P. Kent; bottom right: Steven Wayne Rotsch/DPI; **99** left: Larry Ulrich/DRK Photo; center and right: David Muench Photography Inc.; **100** left: David Muench Photography Inc.; top right: Jack Jeffrey/Photo Resource Hawaii; bottom right: Stephen J. Krasemann/DRK Photo; **101** left: Andy Sacks/Tony Stone Worldwide/Chicago Ltd.; center: David Muench Photography Inc.; right: Jack Parsons/Omni-Photo Communications, Inc.; **102** left: Tony Stone Worldwide/Chicago Ltd.; top right: Wayne Lankinen/DRK Photo; bottom right: George Mars Cassidy/Tony Stone Worldwide/Chicago Ltd.; **103** top left and bottom left: David Muench Photography Inc.; right: Jeff Foott Productions; **104** top left: C. Allan Morgan/DRK Photo; right and bottom left: Stephen J. Krasemann/DRK Photo; **105** top: John Bova/Photo Researchers, Inc.; bottom: Jeff Lepore/Photo Researchers, Inc.; **109** David Muench Photography Inc.; **110** top: Lafayette Long/Goddard Space Flight Center; bottom: Guido Alberto Rossi/Image Bank; **111** Alan J. Posey/Goddard Space Flight Center; **112** San Diego Historical Society, Ticor Collection; **113** Terry Domico/West Stock, Inc.; **114** Luana George/Black Star; **118** top: Howard B. Bluestein/University of Oklahoma/Photo Researchers, Inc.; bottom: Runk/Schoenberger/Grant Heilman Photography; **132** David Muench Photography Inc.; **134** Tony Stone Worldwide/Chicago Ltd.